MATHEMATICS RESEARCH DEVELOPMENTS

MATHEMATICS FOR AGRICULTURAL AND LIFE SCIENCES

PRINCIPLES OF CALCULUS WITH SOLVED PROBLEMS

MATHEMATICS RESEARCH DEVELOPMENTS

MATHEMATICS RESEARCH DEVELOPMENTS

MATHEMATICS FOR AGRICULTURAL AND LIFE SCIENCES

PRINCIPLES OF CALCULUS WITH SOLVED PROBLEMS

CLAUDIO CAPRARA

nova science publishers
New York

Library of Congress Cataloging-in-Publication Data

Names: Caprara, Claudio, author.
Title: Mathematics for agricultural and life sciences : principles of calculus with solved problems / Claudio Caprara.
Description: New York : Nova Science Publishers, [2020] | Series: Mathematics research developments | Includes bibliographical references and index. | Summary: "One of the difficulties that arise in teaching mathematics is related to the identification of the target and the most appropriate teaching methods for the people who are part of it. This aspect, true for all disciplines, applies to mathematics in particular. In fact, for example, an axiomatic approach is certainly suitable for Mathematical, Physical and Engineering Sciences, while students of many applied sciences, such as Agricultural and Life Sciences, need to focus on calculation tools and methodologies useful for their professional development rather than in dealing with the theoretical foundations of mathematics. The peculiarity of this book is not so much in setting classical approach "Theorem: Hypothesis, Thesis" with relative proofs, but in adopting a more pragmatic approach that renounce classical demonstrations, while maintaining a formal coherence in the topics dealt with. In this perspective, considering the approach required by the target to which it is addressed, the objective of this book is to provide methods to studying the variation of a phenomenon and its cumulative effects and consequently the study of the functions and the calculation of integrals respectively. One of the qualifying features is given by a series of completely resolved problems, occupying two-thirds of the volume, in which each mathematical step is detailed to understand "step by step" how to obtain the solution"-- Provided by publisher.
Identifiers: LCCN 2020021667 (print) | LCCN 2020021668 (ebook) | ISBN 9781536180275 (paperback) | ISBN 9781536180282 (adobe pdf)
Subjects: LCSH: Calculus. | Calculus--Problems, exercises, etc.
Classification: LCC QA303.2 .C367 2020 (print) | LCC QA303.2 (ebook) | DDC 515--dc23
LC record available at https://lccn.loc.gov/2020021667
LC ebook record available at https://lccn.loc.gov/2020021668

Published by Nova Science Publishers, Inc. † New York

CONTENTS

PREFACE

This book is based on my multi-years teaching experience in the Degree Courses of Agricultural Sciences and Life Sciences.

The assumption on which it is based is the consideration that the students of the degree programs indicated are not so much interested in the study of mathematics as an end in itself, but as a means able to provide them with calculation tools to solve problems that they will find in their student and professional life.

In this perspective, the peculiarity of the book is not so much in setting classical approach "Theorem: Hypothesis, Thesis" with relative proofs, but in adopting a more pragmatic approach that renounces classical demonstrations, while maintaining a formal coherence in the topics dealt with.

One of the qualifying features is given by a series of completely resolved exercises, in which each mathematical step is detailed to understand "step by step" how to obtain the solution.

The part dedicated to the resolution of the problems is twice the theoretical part and specifically the problems related to the study of functions makes up one third of the book.

The study of functions focused on functions containing the logarithm, exponential and power functions, as they are the ones of greatest interest for the Agricultural and Life Sciences.

I hope that this volume may offer aspects of interest to readers from these disciplines.

Chapter 1

PRINCIPLES OF SET THEORY

1.1. INTRODUCTION AND DEFINITIONS

By mathematical axiom we mean an evident truth in itself, which needs no proof.

In the choice of mathematical axioms for the construction of a mathematical theory, there are above all the needs of logic and practicality that most interest us.

A mathematical proposition can be a statement like:

P: "5 is a prime number"
Q: "n is an even natural number"

The implication symbol is defined as:

$$P \Rightarrow Q$$

It means: "P implies Q"; "If P then Q". For example, we consider the two propositions:

P: "n is a natural number multiple of 6"
Q: "n is an even natural number" (where even number means a multiple of 2).

If P is true, then Q is true, therefore $P \Rightarrow Q$ (a natural number multiple of 6 is even).

Another symbology that is adopted is the double implication symbol:

$$A \Leftrightarrow B$$

It means: “A implies B and B implies A”; “if and only if”

$\exists$ it means: “it exists, they existe”;
$\exists$! it means: “it exists and is unique”

For example:

$\exists$ one natural number ... etc.
$\exists$ two natural number ... etc.

$\forall$ it means: “whatever it is”; “whatever they are”; “for every”
: it means: “such that”

1.2. SET THEORY

The notion of set is taken as primitive, where primitive means that we cannot refer to a simpler proposition.

The set is a grouping of objects, defined as elements of the set, which retain however their individuality.

If we consider:

$A, B, \ldots\ldots$ (sets)
$a, b, \ldots\ldots$ (elements of the set)

If a is an element of the set A, is indicated with:

$a \in A$ it is read: “a belongs to A”

If a is not an element of the set A, is indicated with:

$a \notin A$ it is read: “a does not belongs to A”

Let A and B be two sets, it worths:

$$A = B \Leftrightarrow (x \in A \Leftrightarrow x \in B) \tag{1.1}$$

Two sets A and B are equal if and only if, taken an element x belonging to A, x is also belonging to B and vice versa. So two sets are equal if they have the same elements.

Given two sets, where the elements of the set are placed inside the braces:

$$A = \{a, b\}, \ B = \{b, a\}$$

From the definition of equality between sets (Eq. 1.1), two sets are equal if and only if:

$$x \in A \Leftrightarrow x \in B$$

We see that the two sets contain the same elements *a* and *b*, therefore $A = B$.

Given the sets A and B, we indicate that A is a subset of B with the inclusion symbol:

$$A \subseteq B$$

which means: "A is subset of B"; "A is included in B"; in particular, the symbol more exactly means "included or equal".

The writing means:

$$A \subseteq B \Leftrightarrow (x \in A \Rightarrow x \in B) \tag{1.2}$$

which it states: "if an element *x* belongs to set A, then it also belongs to set B"; this does not exclude that it is A = B.

Instead, it is indicated that A is proper subset of B with:

$$A \subset B$$

which indicates "A is a subset of B and A is different from B":

$$A \subset B \Leftrightarrow A \subseteq B, \ \exists b \in B : b \notin A \tag{1.3}$$

It means: "there exists an element *b* belonging to set B such that *b* does not belong to set A".

An example of a proper subset is shown in Figure 1.1

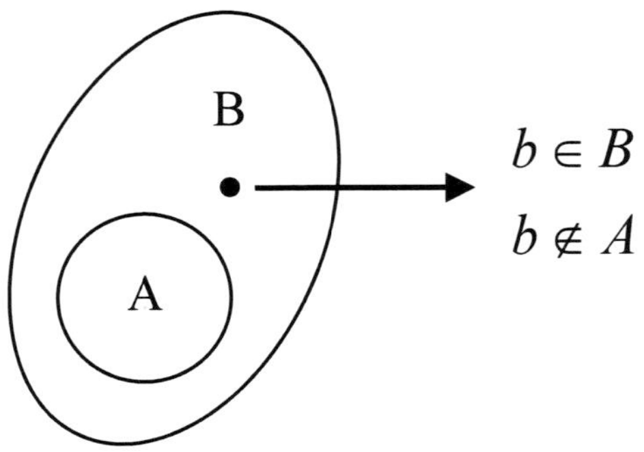

Figure 1.1. Graphical representation of A as a proper subset of B.

Examples of subsets are:

$$\{a,b\} \subseteq \{a,b,c\} \subseteq \{a,b,c,d\}$$

In order to write that:

$$\{a,b\} \subset \{a,b,c\} \text{ (proper subset)}$$

It must be shown that c is not in the first set. For example:

$$\{a,b\} \subset \{a,b,c\} \text{ if } c \neq a \text{ and } c \neq b$$

In the case that the set contains an infinite number of elements, it can be used a form that expresses the property used for the grouping the elements.

For example, the set of natural numbers is the set of numbers 1, 2, 3, ..., etc.:

$$N = \{1,2,3,.........\}$$

If M is the set of even natural numbers:

$$M = \{2,4,6,.........\}$$

the set M is indicated as follows:

$$M = \{n \in \boldsymbol{N} : n\_even\}$$

which means: "M is equal to the set of elements *n* belonging to the set N with the property of being even".

In general, if we have a set A and on A a property *P(x)* of some of its elements, we indicate the elements with the property *P(x)* as:

$$B = \{x \in A : P(x)\} \quad (1.4)$$

B is the set of elements *x* belonging to A for which the property *P(x)* holds.

The *empty set* is a set that has no elements and can therefore be considered a subset of any set.

The empty set is indicated by the symbol ϕ and can be defined for example as:

$$\phi = \{n \in N : n \neq n\} \quad (1.5)$$

which is the set of elements *n* belonging to the set N with the property of being different from themselves, so it is a set without elements (no element has this property), therefore it is the empty set.

Given the sets:

$A, B, \ldots\ldots \subseteq X$ (*A*, *B*, …… subsets of *X*)

It is defined as *union* of set A with set B:

$$A \bigcup B = \{x \in X : x \in A\_or\_x \in B\} \quad (1.6)$$

$A \bigcup B$ is the set of elements *x* belonging to set X that belong to set A or set B, or to both.

It is defined as the *intersection* of set A with set B:

$$A \bigcap B = \{x \in X : x \in A\_and\_x \in B\} \quad (1.7)$$

$A \bigcap B$ is the set of elements *x* belonging to set X that belong to set A and set B.

Given the sets:

$A \subseteq X$

It is defined as *complement* of set A, the set:

$$A' = \{x \in X : x \notin A\} \tag{1.8}$$

A' is the set of elements *x* belonging to set X that do not belong to set A.

A graphic example of the union and intersection of sets and the complement of a set is shown in figures 1.2, 1.3 and 1.4.

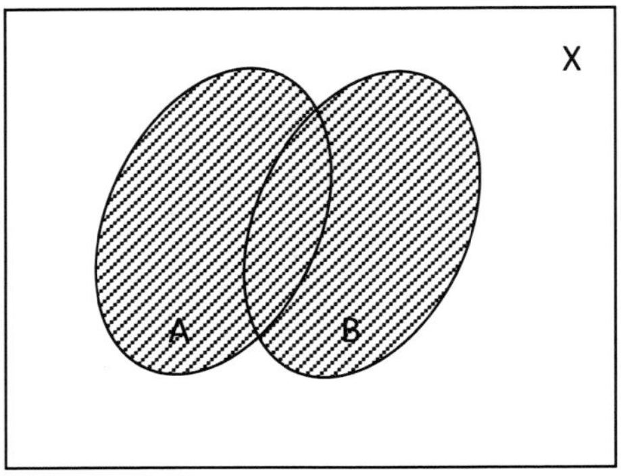

Figure 1.2. Union of set A with set B: $A \cup B$.

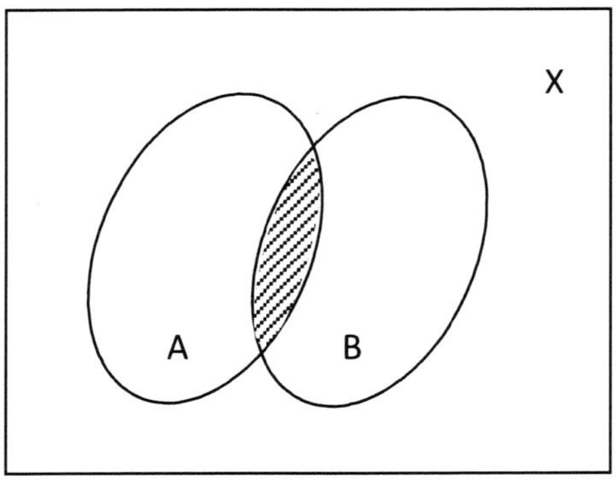

Figure 1.3. Intersection of set A with set B: $A \cap B$.

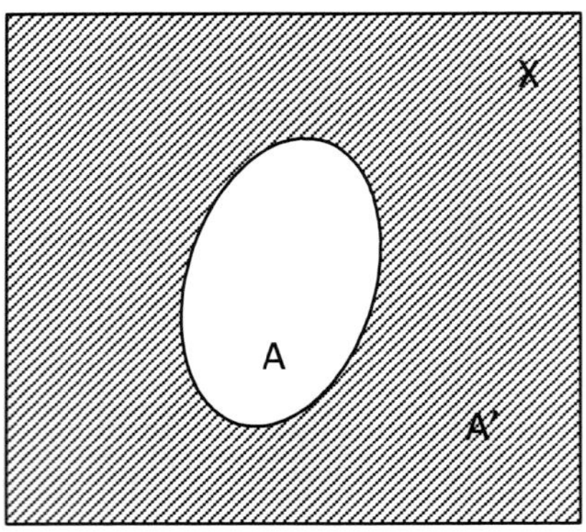

Figure 1.4. Complement of set A.

Some properties apply to the union and intersection of sets.

Idempotency property:

$$A \cup A = A$$
$$A \cap A = A$$

The union of set A with itself is equal to set A itself (similarly for the intersection of set A with itself).

Associative property:

$$A \cup (B \cup C) = (A \cup B) \cup C$$
$$A \cap (B \cap C) = (A \cap B) \cap C$$

Distributive property:

$$A \cup (B \cap C) = (A \cup B) \cap (A \cup C)$$
$$A \cap (B \cup C) = (A \cap B) \cup (A \cap C)$$

Cartesian product.

Given:

A, B sets

(*a, b*) ordered pair

Where:

a = first component or first coordinate

b = second component or second coordinate

Two ordered pairs are equal if:

$$(a,b) = (c,d) \Leftrightarrow \begin{matrix} a = c \\ b = d \end{matrix}$$

if and only if: the 1st coordinate of the first pair is equal to the 1st coordinate of the second pair and the 2nd coordinate of the first pair is equal to the 2nd coordinate of the second pair.

Example:

$(1,2) \neq (2,1)$ the two pairs (1,2) and (2,1) are not equal

It is defined the Cartesian product of A for B (denoted by $A \times B$) the set:

$$A \times B = \{(a,b) : a \in A, b \in B\} \tag{1.9}$$

the set of ordered pairs (*a*, *b*) with the property to be *a* belonging to the set A and *b* belonging to the set B.

Example:

$$A = \{1,2,3\} \quad B = \{1,2\}$$

The Cartesian product of A for B is:

$$A \times B = \{(1,1),(1,2),(2,1),(2,2),(3,1),(3,2)\}$$

The Cartesian product of B for A is:

$$B \times A = \{(1,1),(1,2),(1,3),(2,1),(2,2),(2,3)\}$$

The Cartesian product is not commutative:

$$A \times B \neq B \times A$$

1.3. Notion of Relation

Let A be a set made up of a number of individuals and (a,b) an ordered pair.

For example, we define the kinship relation as:

$$\boldsymbol{P} = \{(a,b) \in A \times A : a\_is\_relative\_of\_b\}$$

The kinship relation is equal to the set of ordered pairs (*a*,*b*), belonging to the Cartesian product of set A for itself, with the property “a is relative of b”.

From the definition of kinship relation we have:

$\mathrm{P} \subseteq A \times A$ the kinship relation is a subset of $A \times A$

or:

$P \subset A \times A$ the kinship relation is a proper subset of $A \times A$

In general, given two sets A and B, $\Re$ is a relation from A to B if it holds:

$$\Re \subseteq A \times B \tag{1.10}$$

$\Re$ is a subset of the Cartesian product $A \times B$.

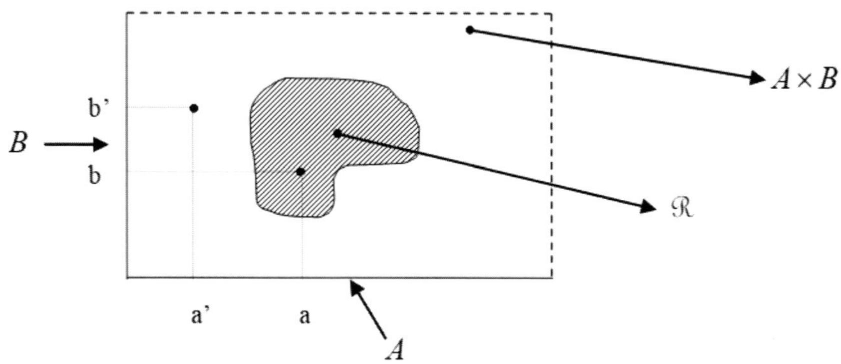

Figure 1.5. Graphical representation of a relation.

From the figure 1.5 it can be seen that the pair (*a*, *b*) is in the relation $\Re$, while the pair (*a*', *b*') is not in the relation $\Re$.

The fact that the pair (*a*, *b*) is in the relation $\Re$ is indicated by:

$(a,b) \in \Re$ which means "the pair (*a*, *b*) belongs to the relation $\Re$";
$a \, \Re \, b$ which means "*a* is in the relation $\Re$ with *b*", or "*b* corresponds to *a* through $\Re$".

While the fact that the pair (*a*, *b*) is not in the relation $\Re$ is indicated by:

$(a',b') \notin \Re$ which means "the pair (*a*, *b*) does not belong to the relation $\Re$";
$a' \not\Re \, b'$ which means "*a* is not in the relation $\Re$ with *b*", or " *b* does not correspond to *a* through $\Re$".

Let ***R*** be the set of real numbers, we define the Cartesian product of ***R*** with itself:

$\boldsymbol{R} \times \boldsymbol{R} = \boldsymbol{R}^2$

By associating $\boldsymbol{R}$ with each of the two orthogonal axes *x*, *y*, we obtain a system of Cartesian axes (Figure 1.6):

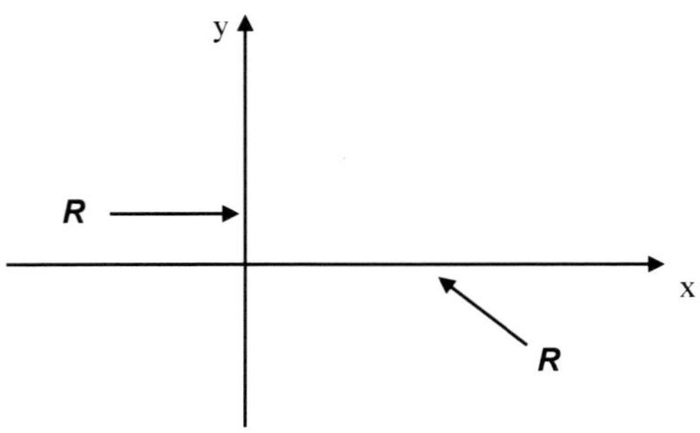

Figure 1.6. Cartesian axis system.

If we consider the bisector of two orthogonal axes, we obtain the graph in the Figure 1.7.

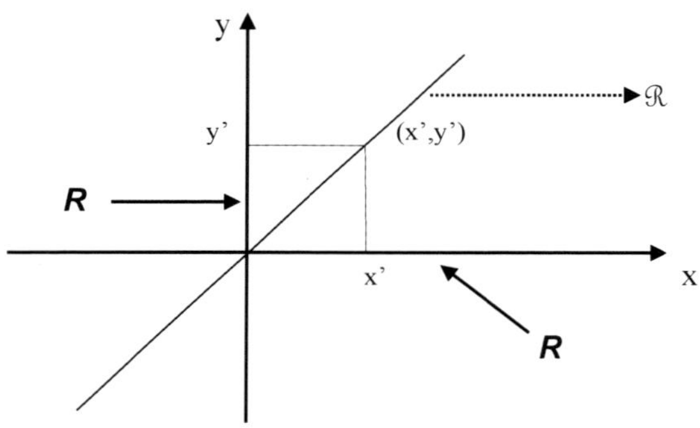

Figure 1.7. Bisector of two orthogonal axes.

The bisector of the two orthogonal axes is defined as a relation $\Re$ if:

$$x \, \Re \, y \quad \Leftrightarrow \; x = y$$

that is, *x* is in the relation $\Re$ with *y* if and only if *x* is equal to *y*.

The relation $\Re$ is defined as:

$$\Re = \{(x, y) \in \boldsymbol{R}^2 : x = y\} \tag{1.11}$$

the relation $\Re$ is equal to the set of ordered pairs *(x, y)*, belonging to $\boldsymbol{R}^2$ such that $x = y$.

There are some properties for relations, which are not valid for all relations.

Given the set *A* and the relation:

$$\Re \subseteq A \times A$$

we define:

Reflexive relation

The relation $\Re$ is reflexive if and only if the pair (a, a) belongs to the relation $\Re$ for every element a belonging to the set A:

$$a \,\Re\, a, \forall a \in A \;\Rightarrow\; \Re \text{ is reflexive.}$$

Symmetric relation

The relation $\Re$ is symmetric if:

$$a \,\Re\, b \Rightarrow \quad b \,\Re\, a$$

It means: "if a is in the relation $\Re$ with b, then b is in the relation $\Re$ with a". If this holds, then the relation $\Re$ is symmetric.

Transitive relation

The relation $\Re$ is transitive if:

$$a \,\Re\, b,\, b \,\Re\, c \Rightarrow a \,\Re\, c$$

It means: "if a is in the relation $\Re$ with b and b is in the relation $\Re$ with a, then a is in the relation $\Re$ with c". If this holds, then the relation $\Re$ is transitive.

Given the set A and the relation:

$$\Re \subseteq A \times A$$

we define:

Partial ordering relation:

$\Re$ is partial ordering relation $\Leftrightarrow$ $\Re$ is reflexive and transitive relation

Total ordering relation:

$\Re$ is total ordering relation $\Leftrightarrow$

$\Re$ is partial ordering relation, $\forall a,b \in A$ it is: $a \,\Re\, b$ or $b \,\Re\, a$.

Equivalence relation:

$\Re$ is equivalence relation $\Leftrightarrow$ $\Re$ is reflexive, symmetric and transitive relation.

An example of equivalence is given by parallel lines in the plane.

A line is always parallel to itself (reflexive relation).

If a line is parallel to another line, this other line is parallel to the first one (symmetric relation).

If a first line is parallel to a second line and this second one is parallel to a third line, then the first one is parallel to the third (transitive relation).

Another example of equivalence is the similarity transformation of triangles.

The relation of $\geq$ for real numbers is reflective and transitive:

Each number is equal to itself.

If a $\geq$ b and b $\geq$ c, then a $\geq$ c.

The relation of > for real numbers is not reflexive but only transitive:

No number is greater than itself.

If a > b and b > c, then a > c.

The relation of $\geq$ for real numbers is total ordering relation:

Each number is equal to itself (reflexive relation).

If a $\geq$ b and b $\geq$ c, then a $\geq$ c (transitive relation)

$\forall a, b \in \boldsymbol{R}$ it is: a $\geq$ b or b $\geq$ a.

Chapter 2

REAL NUMBERS

2.1. SET OF REAL NUMBERS

An *algebraic structure* is defined as formed by one set and a number of operations and relations defined on that set.

The *set of real numbers* $\boldsymbol{R}$ constitutes an algebraic structure and is defined by a certain number of properties, divided into four groups.

The first group of properties is obtained by defining in $\boldsymbol{R}$ the *addition* operation and it has the following properties:

$$(\boldsymbol{R},+)\begin{cases} a+b=b+a \quad ,\forall a,b\in \boldsymbol{R} \\ a+(b+c)=(a+b)+c \quad ,\forall a,b,c\in \boldsymbol{R} \\ \\ \exists! \;\; 0\in \boldsymbol{R}: a+0=a \;\; ,\forall a\in \boldsymbol{R} \\ \forall a\in \boldsymbol{R}, \quad \exists(-a)\in \boldsymbol{R}: \quad a+(-a)=0 \end{cases} \tag{2.1}$$

The second group of properties is obtained by defining in $\boldsymbol{R}$ the *multiplication* operation:

$$(\boldsymbol{R},\cdot)\begin{cases} a\cdot b=b\cdot a \quad ,\forall a,b\in \boldsymbol{R} \\ a\cdot(b\cdot c)=(a\cdot b)\cdot c, \forall a,b,c\in \boldsymbol{R} \\ \exists! \;\; 1\in \boldsymbol{R}, 1\neq 0: a\cdot 1=a \;\; ,\forall a\in \boldsymbol{R} \\ \forall a\in \boldsymbol{R}\text{-}\{0\} \quad \exists a^{-1}\in \boldsymbol{R}: \quad a\cdot a^{-1}=1 \\ \end{cases} \tag{2.2}$$

$$a \cdot (b + c) = a \cdot b + a \cdot c \, , \ \forall a, b, c \in \boldsymbol{R}$$

The third group of properties is obtained by defining in ***R*** a *total ordering relation* $\leq$:

$$(\boldsymbol{R}; \leq) \quad \begin{cases} a \leq b & , c \in \boldsymbol{R} & \Rightarrow & a + c \leq b + c \\ \\ a \leq b & , c \in \boldsymbol{R}, c \geq 0 & \Rightarrow & a \cdot c \leq b \cdot c \end{cases} \tag{2.3}$$

The fourth group of properties is formed by the *completeness property* of ***R*** with respect to the relation $\leq$.

To define the completeness property, a definition of *contiguous classes* must first be given.

Let be given $A, B \subseteq \boldsymbol{R}$, it is stated that:

$A, B \subseteq \boldsymbol{R}$ are contiguous classes (indicated with c.c.) $\Leftrightarrow$

$$a \in A, b \in B \quad \Rightarrow \quad a \leq b \tag{2.4}$$

Graphically, two sets that have this property are represented in Figure 2.1.

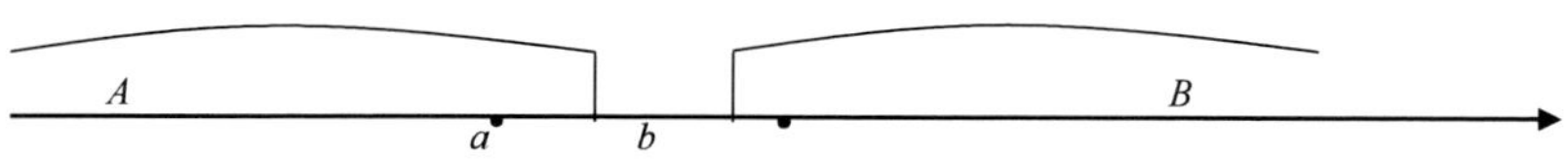

Figure 2.1. Graphical representation of two contiguous classes.

For example:

$$A =]-\infty, 0[= \{x \in \boldsymbol{R} : x < 0\}$$
$$B =]0, +\infty[= \{x \in \boldsymbol{R} : x > 0\}$$

Then:

$$a \in A \Leftrightarrow a < 0$$
$$b \in B \Leftrightarrow b > 0$$

We have therefore:

$a < 0, 0 < b \Rightarrow a < b$

Two subsets of R that have this property are called contiguous classes (c.c.):

$A, B \subseteq \boldsymbol{R} \quad , a \in A, b \in B \Rightarrow a \leq b$ (*A, B* are c.c.)

We define the completeness property of $(\boldsymbol{R}, \leq)$:

$$A, B \text{ are c.c.}, \Rightarrow \quad \exists \lambda \in \boldsymbol{R} : a \leq \lambda, \lambda \leq b, \forall a \in A, \forall b \in B \tag{2.5}$$

If A and B are contiguous classes, there exists an element λ belonging to ***R*** such that $a \leq \lambda$ and $\lambda \leq b$ for every $a \in A$ and for every $b \in B$.

λ is called the separation element, in the example considered it is λ = 0.

This property is called the completeness property of ***R*** with respect to the relation $\leq$.

2.2. SETS OF NATURAL, INTEGER, RATIONAL AND IRRATIONAL NUMBERS

The set of natural numbers is constituted by the numbers that are obtained by gradually adding 1 to the unit, it consists of the set of positive integers excluding 0:

$$N = \{1,2,3,4,5,\ldots\}: \qquad 1 \in N, n \in N \Rightarrow n+1 \in N \tag{2.6}$$

$n = 1 + 1$

$n' = n + 1$

$n'' = n' + 1$

....

We define the set of integers numbers, indicated with *Z*:

$$Z = \{m - n : m, n \in N\} \tag{2.7}$$

We define the set of rational numbers, indicated with *Q*:

$$Q = \left\{ \frac{m}{n} : m \in I, n \in N \right\} \tag{2.8}$$

Where:

$N \subset I \subset Q \subset \boldsymbol{R}$

We define the set of irrational numbers:

$$\boldsymbol{R} - Q \tag{2.9}$$

It is the set of numbers that cannot be obtained as a ratio between two integers. Q does not have the completeness property, for example:

$\sqrt{2} \notin Q$ it is not a rational number

2.3. Limit Superior and Limit Inferior of a Set

2.3.1. Upper Bound of a Set

Before defining the limit superior of a set, the *upper bound of a set* is defined. Let:

$A \subseteq \boldsymbol{R} \quad , A \neq \phi$

$m_1 \in \boldsymbol{R}$ is an upper bound of A if: $m_1 \geq a \quad , \forall a \in A$

Examples:

$A =]0,1[= \{x \in \boldsymbol{R} : 0 < x < 1\}$

In this case, for example:

$m_1 = 2$, $m_1 = 3$ are upper bounds of A.

In the case of:

$A =]0,+\infty[$ A has no upper bounds.

A set is superiorly bounded if it has upper bounds:

$$A \text{ is superiorly bounded} \Leftrightarrow \exists m_1 \in \boldsymbol{R} : m_1 \geq a, \forall a \in A \tag{2.10}$$

2.3.2. Limit Superior of a Set

Let:

$$A \subseteq \boldsymbol{R} \qquad ,A \neq \phi$$

We define λ *limit superior* of *A*, or *least upper bound* of *A*:

$$\lambda \in \boldsymbol{R} \qquad ,\lambda = \sup A \text{ if:}$$
$$\lambda \geq a \qquad ,\forall a \in A$$
$$m_1 \geq a \qquad ,\forall a \in A \qquad \Rightarrow m_1 \geq \lambda \tag{2.11}$$

That is, λ must be an upper bound of A and among all upper bounds λ must be the smallest. The limit superior is unique.

2.3.3. Lower Bound of a Set

Similarly, before defining the limit inferior of a set, the *lower bound of a set* is defined. Let:

$$A \subseteq \boldsymbol{R} \qquad ,A \neq \phi$$

$m_2 \in \boldsymbol{R}$ is a lower bound of *A* if: $m_2 \leq a \qquad ,\forall a \in A$

A set is inferiorly bounded if it has lower bounds:

$$A \text{ is inferiorly bounded} \Leftrightarrow \exists m_2 \in \boldsymbol{R} : m_2 \leq a, \forall a \in A \tag{2.12}$$

2.3.4. Limit Inferior of a Set

Let:

$$A \subseteq \boldsymbol{R} \qquad ,A \neq \phi$$

We define μ *limit inferior* of *A*, or *greatest lower bound* of *A*:

$$\mu \in \boldsymbol{R} \qquad , \mu = \inf A \text{ if:}$$
$$\mu \leq a \qquad , \forall a \in A$$
$$m_2 \leq a \qquad , \forall a \in A \qquad \Rightarrow m_2 \leq \mu \tag{2.13}$$

That is, μ must be a lower bound of A and among all lower bounds μ must be the greatest. The limit inferior is unique.

2.3.5. Maximum and Minimum of a Set

$A \subseteq \boldsymbol{R}$ has *maximum*

if and only if:

$$\exists \sup A \qquad , \sup A \in A \tag{2.14}$$

The maximum is the limit superior when this belongs to the set (otherwise the set has no maximum). If the set has maximum:

$$\max A = \sup A$$

Examples:

$$A =]0,1[\qquad 1 = \sup A, 1 \notin A \qquad \Rightarrow A \text{ has no maximum}$$
$$A =]0,1] \qquad 1 = \sup A, 1 \in A \qquad \Rightarrow A \text{ has maximum}$$

$A \subseteq \boldsymbol{R}$ has *minimum*

if and only if:

$$\exists \inf A \qquad , \inf A \in A \tag{2.15}$$

The minimum is the limit inferior when this belongs to the set (otherwise the set has no minimum). If the set has minimum:

$$\min A = \inf A$$

Examples:

$A =]0,1[\qquad 0 = \inf A, 0 \notin A \qquad \Rightarrow A$ has no minimum

$A = [0,1[\qquad 0 = \inf A, 0 \in A \qquad \Rightarrow A$ has minimum

Chapter 3

FUNCTIONS OF REAL VARIABLE

3.1. DEFINITION OF FUNCTION

Given two sets A and B, we define:

f is a function from A to B $\Leftrightarrow$
f is a relation from A a B: $\forall a \in A \ \exists! b \in B : (a,b) \in f$ (3.1)

f is a function from A to B if and only if f is a relation from A to B such that, for every element a belonging to the set A, there exists and is unique an element b belonging B such that the pair (a, b) belongs to f.

A function is therefore a relation that, to an element of the 1st set, corresponds one and only one element of the 2nd set.

Some graphic examples are shown in Figures 3.1, 3.2, and 3.3.

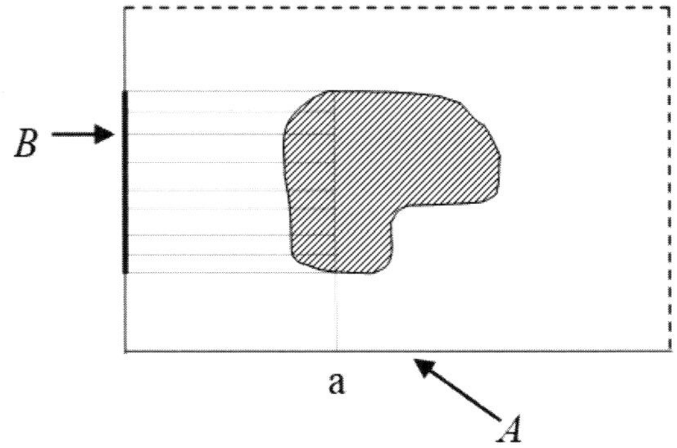

Figure 3.1. It is not a function, at one point of the set A there correspond multiple elements of the set B.

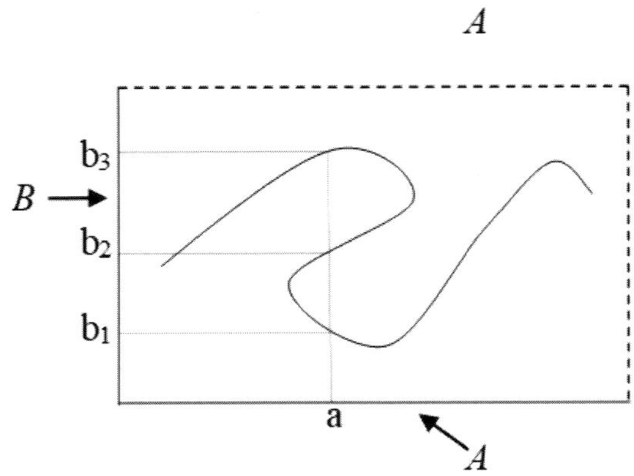

Figure 3.2. It is not a function, at one point of the set A there correspond some elements of the set B.

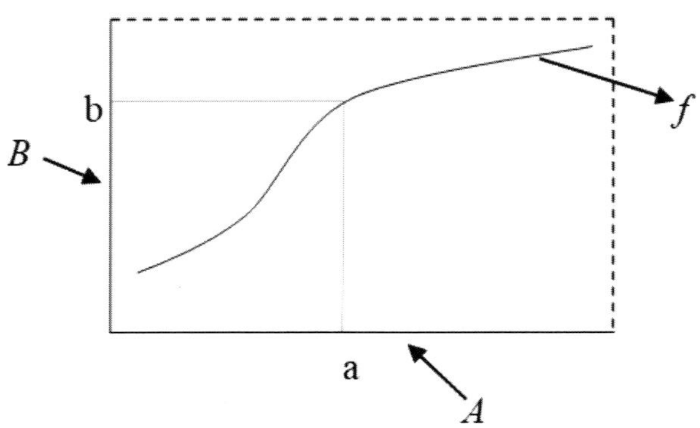

Figure 3.3. It is a function, one point in the set A corresponds to one element of the set B.

So, given two sets *A, B, t*he conditions for *f* to be a function are:

$$\begin{aligned} &\text{i)} \quad f \subseteq A \times B \\ &\text{ii)} \quad \forall a \in A, \exists! b \in B : (a,b) \in f \end{aligned} \tag{3.2}$$

that is, that *f* is a relation and that, for every element *a* belonging to A, there exists and is unique an element *b* belonging to B such that the pair (*a*, *b*) belongs to the relation *f*.

To indicate the function *f* from A to B, we use the notation:

$$\begin{aligned} &f : A \rightarrow B \\ &(a,b) \in f \equiv b = f(a) \end{aligned} \tag{3.3}$$

The notation $(a,b) \in f$ coincides with the notation *b* = *f(a)*, where *b* = *f(a)* means that *b* is the image through *f* of *a*.

A schematic representation of function is shown in Figure 3.4.

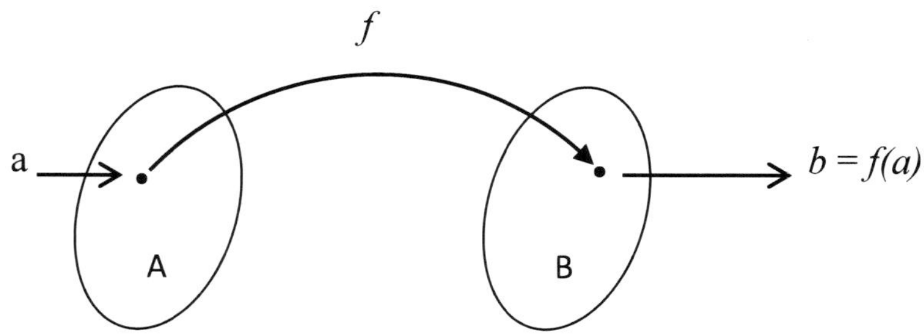

Figure 3.4. Graphical scheme of a function based on set theory.

$$f : A \to B$$

where A is the domain of *f* and B is the set to which the values of *f* belong.

We define codomain:

$$\mathcal{C}(f) = \{b \in B : \exists a \in A : b = f(a)\} \tag{3.4}$$

The codomain is the set of elements *b* belonging to B, with the property that there exists an element *a* belonging to A such that *b* is the image through *f* of *a* (expressed by $b = f(a)$).

The codomain, therefore, is the totality of the elements of B that have a point in A through the function *f*.

3.2. Surjection, Injection and Bijection of a Function

$$f \text{ is } \textit{surjective} \Leftrightarrow \mathcal{C}(f) = B \tag{3.5}$$

f is surjective if and only if the codomain of *f* coincides with B.

$$f \text{ is } \textit{injective} \Leftrightarrow \left(f(a') = f(a'') \Rightarrow a' = a''\right) \tag{3.6}$$

f is injective if and only if: if we have that the image through *f* of *a*' is equal to the image through *f* of *a*'', then *a*' is equal to *a*''. A function can be said to be injective if it brings distinct points to distinct points.

$$f \text{ is a } \textit{bijection} \Leftrightarrow f \text{ is surjective and injective} \tag{3.7}$$

Surjection, injection and bijection are indicated respectively with:

$$f : A \xrightarrow[su]{} B$$

$$f : A \xrightarrow{1-1} B$$

$$f : A \xrightarrow[su]{1-1} B$$

3.3. Examples of Functions

Let:

$$f: \boldsymbol{R} \to \boldsymbol{R},\ y = ax + b$$
$$f = \{(x, y) \in \boldsymbol{R}^2 : y = ax + b\}\ a, b \in \boldsymbol{R} \tag{3.8}$$

We want to see if *f* is a function, that is, we need to check whether:

$$\forall x \in \boldsymbol{R}, \exists! y \in \boldsymbol{R} : (x, y) \in f$$

A straight line is obtained, the coordinates of which when intercepting the axes are obtained by setting x and y equal to 0 in the equations:

$$y = ax + b$$
$$x = \frac{y - b}{a}$$

Setting equal to 0 we obtain for the two formulas:

$$x = 0 \quad \to \quad y = b$$

$$y = 0 \quad \to \quad x = -\frac{b}{a}, \text{ with } a \neq 0$$

This is true, as can also be seen from the graph in Figure 3.5:

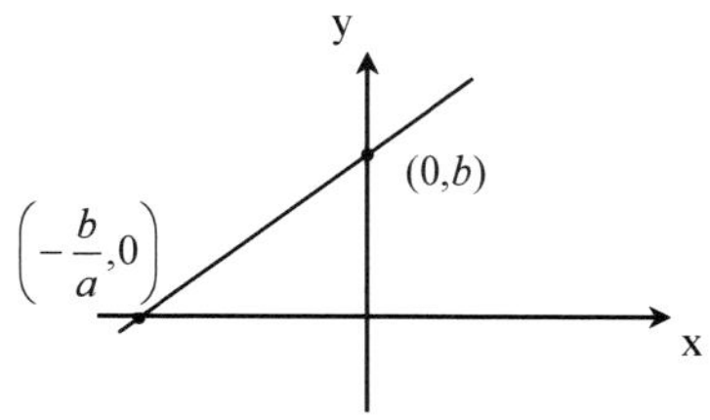

Figure 3.5. Graph of the function representing a straight line.

The above graph was obtained with $a \neq 0$.
With $a = 0$, the graph would be a horizontal straight line (Figure 3.6):

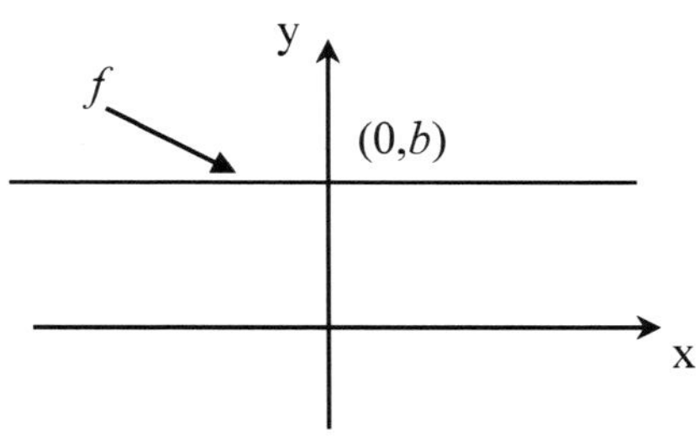

Figure 3.6. Graph of a function representing a straight line with $a = 0$.

We look for the codomain of the function. By definition, it is:

$$\mathcal{C}(f) = \{y \in \boldsymbol{R} : \exists x \in \boldsymbol{R} : y = ax + b\} = \boldsymbol{R} \ (a \neq 0)$$
$$\mathcal{C}(f) = \{y \in \boldsymbol{R} : y = b\} = \{b\} \ (a = 0)$$

We look at whether the function is surjective.

$$f : \boldsymbol{R} \to \boldsymbol{R}$$

f is surjective if its codomain coincides with $\boldsymbol{R}$.
There are two cases:

$$a \neq 0$$
$$a = 0$$

First case ($a \neq 0$):

$\mathcal{C}(f) = \boldsymbol{R}$ (therefore, with $a \neq 0$, *f* is *su*)

Second case ($a = 0$):

$\mathcal{C}(f) = \{b\} \neq \boldsymbol{R}$ (with $a = 0$, *f* is not *su*)

From the definition of injection, *f* is 1-1 if:

$$f(x') = f(x'') \Rightarrow x' = x''$$

First case ($a \neq 0$):

$ax' + b = ax'' + b \Rightarrow x' = x''$ (therefore, with $a \neq 0$, *f* is *1-1*)

Second case ($a = 0$):

It is clearly seen that in this case *f* is not *1-1*. In fact, the image through *f* of all *x* (*x'*, *x''*, *x'''*, ... etc.) is always the same, even if the *x* are not equal.

We look at whether the function is a bijection.

In the case where $a \neq 0$, *f* is *su* and *1-1*, for which *f* is a bijection.

In the case where $a = 0$, *f* is neither *su* nor *1-1*, for which *f* is not a bijection.

Let's consider another example of a function:

$$f: \boldsymbol{R} \to \boldsymbol{R},\ y = x^2$$
$$f = \{(x, y) \in \boldsymbol{R}^2 : y = x^2\} \qquad (3.9)$$

We want to see if *f* is a function, that is if:

$$\forall x \in \boldsymbol{R}, \exists!\, y \in \boldsymbol{R} : (x, y) \in f, y = x^2$$

This is true, as can also be seen from the graph in Figure 3.7.

We look at whether the function is surjective.

$$f : \boldsymbol{R} \to \boldsymbol{R}$$

f is surjective if its codomain coincides with ***R***.

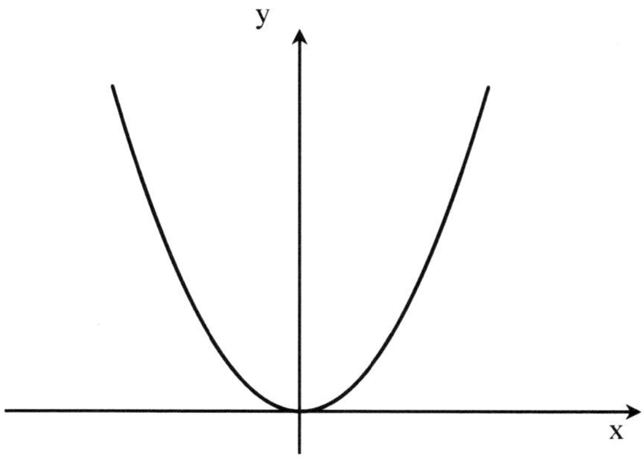

Figure 3.7. Graphical representation of the function 3.9, $y = x^2$.

$$\mathcal{C}(f) = \{y \in \boldsymbol{R} : \exists x \in \boldsymbol{R} : y = x^2\}$$

$$y = x^2 = (-x)^2 \geq 0$$

$\mathcal{C}(f) = \{y \in \boldsymbol{R} : y \geq 0\} = [0, +\infty[$ the function is not surjective

The symbol "[" on the left means "closed" (the interval contains the 0) while the symbol "[" on the right means "open" (the interval subset of real numbers does not contain $+\infty$).

In general:

"["left closed interval,"]" right closed interval.
"]"left open interval," [" right open interval.

We look at whether the function is injective.
From the definition of injection, *f* is 1-1 if:

$$f(x') = f(x'') \Rightarrow x' = x''$$

By setting:

$$x' = x \quad , \quad x'' = -x$$

We have:

$$f(x') = x^2 \quad , \quad f(x'') = (-x)^2 = x^2 \quad \Rightarrow \quad x' \neq x''$$

The function is not *1-1*.

3.4. COMPOSITE FUNCTION

Let:

A, B, C sets,
$f: A \to B$, $g: B \to C$

we define the *composite function h* and we indicate it with the notation:

$$h = g \circ f$$
$$h = \{(a,c) \in A \times C : \exists b \in B : b = f(a); c = g(b)\} \quad (3.10)$$

The composite function *h* is equal to the set of ordered pairs (*a*, *c*), belonging to the Cartesian product $A \times C$, with the property that there exists an element *b* belonging to B such that *b* is the image through *f* of *a* and that *c* is the image through *g* of *b*.

The notation:

$h = g \circ f$ it is read: "*f* composed with *g*" (that is, we write to the right the function that is applied first).

c is the image through *h* (composite function) of *a*:

$c = h(a) \Leftrightarrow c = g(f(a)) = (g \circ f)a$

it is read: "*f* composed with *g*" applied to point *a*"

A graphical example of a composite function is shown in Figure 3.8.

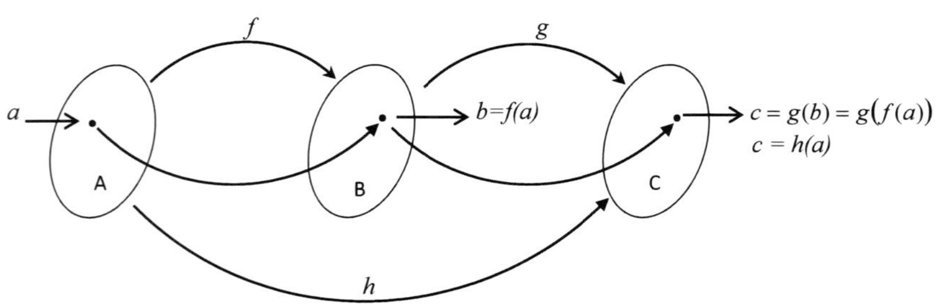

Figure 3.8. Graphical representation of a composite function.

So in summary:

A, B, C sets,
$f: A \to B$, $g: B \to C$

We define a composite function:

$$h = g \circ f \quad h : A \to C\,,\ h(a) = g\big(f(a)\big), \forall a \in A \tag{3.11}$$

Consider for example:

$$f : \boldsymbol{R} \to [0,+\infty[\quad , \quad f(x) = x^2 \text{ (the function is } su\text{)} \tag{3.12}$$

$$g : [0,+\infty[\to \boldsymbol{R} \quad , \quad g(y) = \sqrt{y} \tag{3.13}$$

The composite function:

$$h = g \circ f$$

can be graphically schematized as in Figure 3.9.

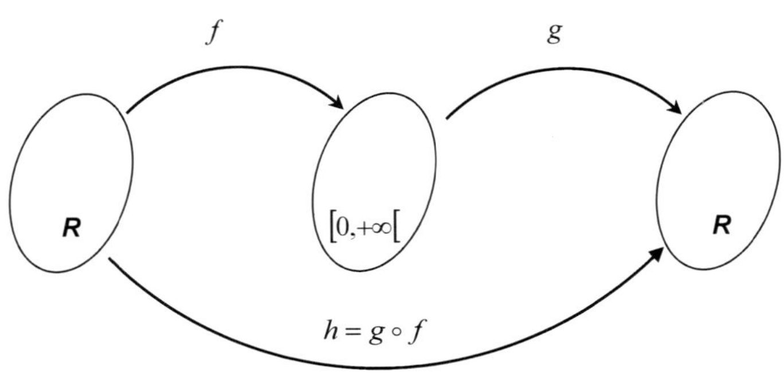

Figure 3.9. Graphical representation of functions 3.11, 3.12 and 3.13.

The composite function is:

$$h : \boldsymbol{R} \to \boldsymbol{R}$$
$$h(x) = g\big(f(x)\big) = g\big(x^2\big) = \sqrt{x^2}$$

3.5. Inverse Function

Let:

A, B sets,
$f : A \to B$

we define the *inverse function*:

$$g : B \rightarrow A$$

A graphical example of an inverse function is shown in Figure 3.10.

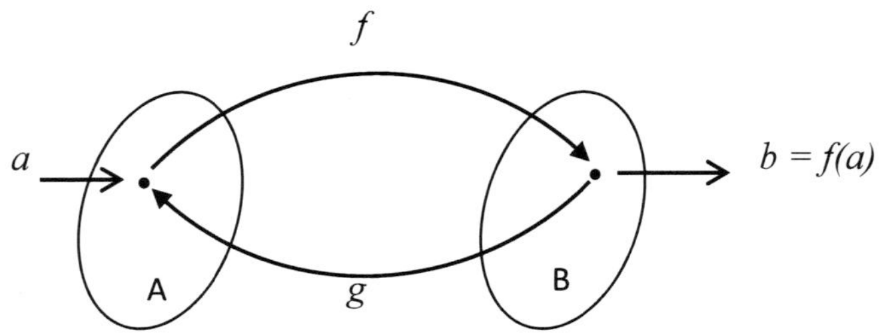

Figure 3.10. Graphical representation of inverse function.

If there is an inverse function g, then f is invertible.
This means that:

A, B sets
$f : A \rightarrow B$ is invertible $\Leftrightarrow \exists g : B \rightarrow A,\ a = g(b)$

so that:

$$g(f(a)) = a \qquad ,\forall a \in A$$
$$f(g(b)) = b \qquad ,\forall b \in B$$

The image through g of $f(a)$ is equal to a, while the image through f of $g(b)$ is equal to b, for every a and b.

The inverse function of a function f is indicated by the notation:

$$g = f^{-1} \tag{3.14}$$

In equation 3.14 f is a function, not to be confused with a variable x.
For example, the function whose graph is represented in Figure 3.11 it is not invertible.
To be invertible, a function must be *1-1* and *su* (without proof).
As an example, the square and the square root are inverse functions (one is the inverse function of the other).

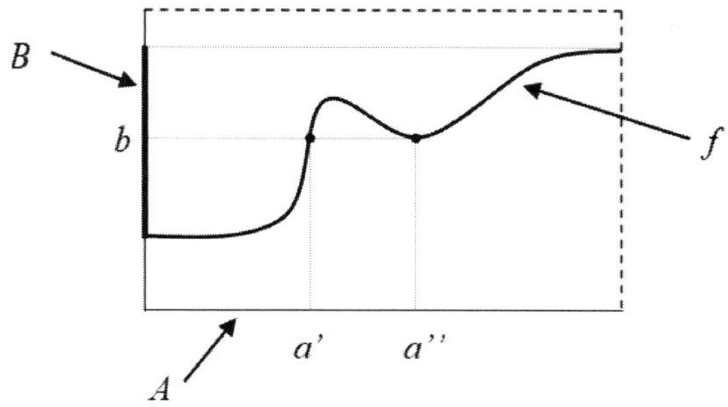

Figure 3.11. Graphical example of not invertible function.

In fact, they verify:

$$x^2 = y \qquad \Leftrightarrow \qquad \sqrt{y} = x$$

That is, the square of x is equal to y if and only if the square root of y is equal to x. However, this only applies to:

$$x \geq 0$$

In this sense, the previous example of composite function (Eq. 3.12):

$$f: \boldsymbol{R} \rightarrow [0,+\infty[,\ f(x) = x^2$$

cannot be used as an example of inverse function, but must be:

$$f: [0,+\infty[\rightarrow [0,+\infty[,\ f(x) = x^2 \qquad (3.15)$$

3.6. Some Considerations on the Domain of a Function

If we consider a function of a real variable, it cannot always be indicated with:

$$f\!: \boldsymbol{R} \rightarrow \boldsymbol{R}$$

Its domain does not necessarily coincide with ***R***.

The domain of a function represents the set of points for which the function can be defined, and it is not always true that this set coincides with ***R***.

For example:

$$f(x) = \sqrt{x}$$

It can only be defined for $x \geq 0$ (the square root cannot be calculated for negative real numbers), for which:

$$f : [0,+\infty[\rightarrow \boldsymbol{R}$$

The domain is $[0,+\infty[$.

Considering another example, the function:

$$f(x) = \frac{1}{x}$$

it can be defined for all real values except $x = 0$, because the ratio $\frac{1}{0}$ would give a result tending to $+\infty$ or $-\infty$ ($+\infty$ and $-\infty$ do not belong to the set of real numbers $\boldsymbol{R}$).

The function can be defined for:

$$f\!: \boldsymbol{R}\text{-}\{0\} \rightarrow \boldsymbol{R}$$

The domain is $\boldsymbol{R}\text{-}\{0\}$.

3.7. Absolute Value

Let:

$$x \in \boldsymbol{R}$$

Let's define the *absolute value* of x:

$$|x| = \begin{matrix} x & ,x \geq 0 \\ -x & ,x \leq 0 \end{matrix} \tag{3.16}$$

The absolute value of x is equal to x if x is positive, is equal to $-x$ if x is negative.

We can define the absolute value function:

$$f : \boldsymbol{R} \rightarrow \boldsymbol{R} \qquad , f(x) = |x|$$

The graph of the absolute value function is shown in Figure 3.12.

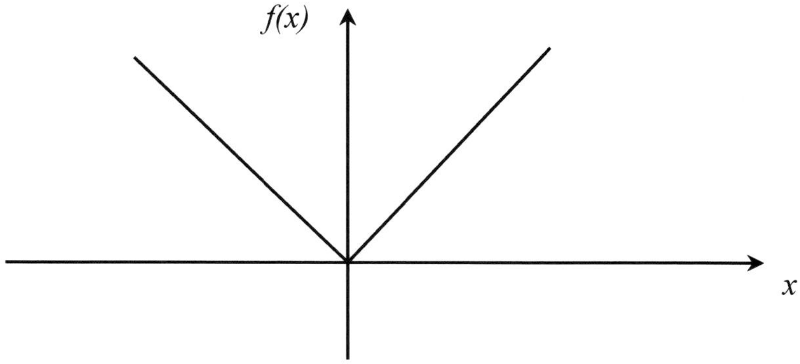

Figure 3.12. Graphical representation of absolute value function.

The following properties apply to the absolute value:

i) $|x| \geq 0 \qquad ,= 0 \Leftrightarrow x = 0$ (3.17a)

ii) $|x| \geq x \qquad , \forall x \in \boldsymbol{R}$ (3.17b)

iii) $|xy| = |x| \cdot |y| \qquad , \forall x, y \in \boldsymbol{R}$ (3.17c)

iv) $|x + y| \leq |x| + |y| \qquad , \forall x, y \in \boldsymbol{R}$ (3.17d)

v) $|x + y| \geq |x| - |y| \qquad , \forall x, y \in \boldsymbol{R}$ (3.17e)

Chapter 4

LIMIT OF A FUNCTION

4.1. NEIGHBORHOOD OF A POINT

Let $x_0 \in \boldsymbol{R}$ be given, we define the *circular neighborhood* of x_0 and radius r as the open set ($r > 0$):

$$S(x_0, r) = \{x \in \boldsymbol{R} : |x - x_0| < r\} \tag{4.1}$$

The neighborhood is an open interval centered in x_0 of amplitude $2r$, (the extremes are excluded, Figure 4.1).

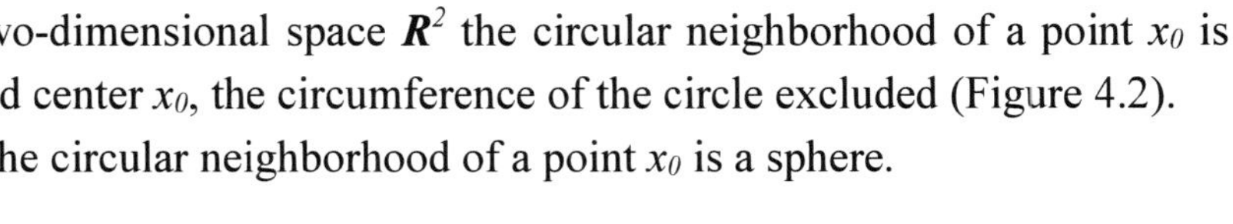

Figure 4.1. Example of circular neighborhood on the set of real numbers.

In a two-dimensional space $\boldsymbol{R}^2$ the circular neighborhood of a point x_0 is a circle of radius r and center x_0, the circumference of the circle excluded (Figure 4.2).

In $\boldsymbol{R}^3$ the circular neighborhood of a point x_0 is a sphere.

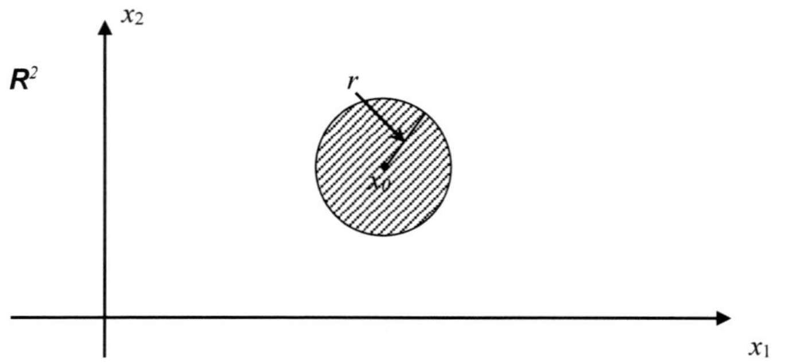

Figure 4.2. Example of circular neighborhood in $\boldsymbol{R}^2$.

4.2. Accumulation Point

We define:

x_0 accumulation point (or *cluster point*) of A, $A \subseteq \boldsymbol{R}$ $\Leftrightarrow$

$$\forall \varepsilon > 0 \qquad S(x_0, \varepsilon) \cap A \text{ is infinite} \tag{4.2}$$

That is, if the intersection of the neighborhood of x_0 with the set A is an infinite set (it contains *infinite points*).

Let's consider the set:

$$A' = A + \{y_0\} \qquad , A \subseteq \boldsymbol{R}$$

Figure 4.3 represents a graphic example of accumulation points in a two-dimensional space $\boldsymbol{R}^2$:

Figure 4.4 represents a graphic example of accumulation points for the set of real numbers $\boldsymbol{R}$.

We define *derived set* of A, denoted by $D(A)$, the set of accumulation points of A:

$$x_0 \in D(A) \qquad \Leftrightarrow x_0 \text{ is an accumulation point of } A. \tag{4.3}$$

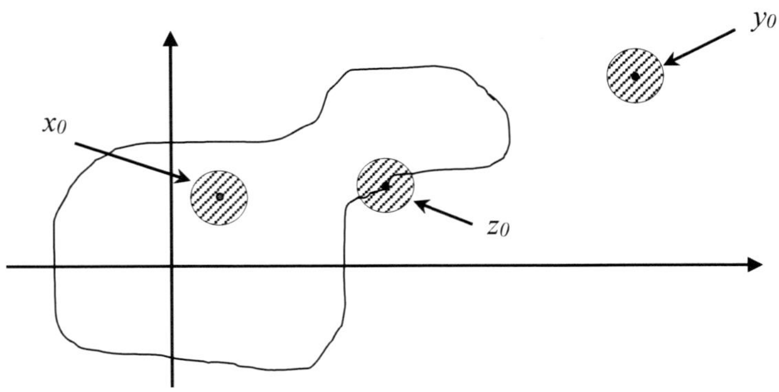

Figure 4.3. Graphic example in $\boldsymbol{R}^2$: x_0 is an accumulation point of A', y_0 is not an accumulation point of A', z_0 is an accumulation point of A'.

Equation 4.3 is a symbolic way of representing the set of accumulation points, "useful" for limit considerations.

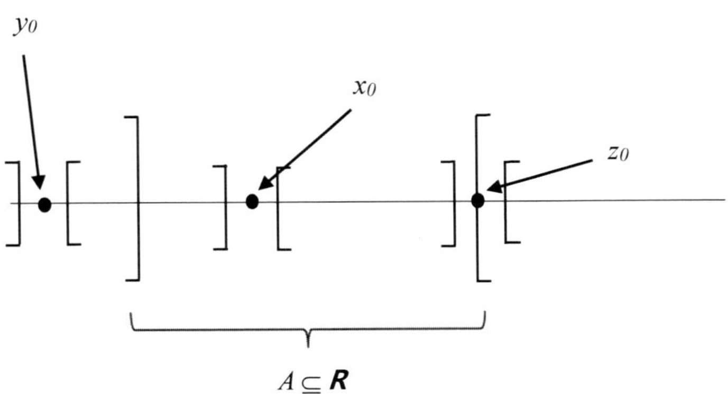

Figure 4.4. Graphic example in $\boldsymbol{R}$: x_0 is an accumulation point of A', y_0 is not an accumulation point of A', z_0 is an accumulation point of A'.

We can make some considerations on $+\infty$ and $-\infty$.

Even if $+\infty, -\infty \notin \boldsymbol{R}$, it is necessary to define neighborhoods of $+\infty$ and $-\infty$.

We define:

$$]a,+\infty[= \{x \in \boldsymbol{R} : x > a\} \tag{4.4}$$

$$]-\infty, b[= \{x \in \boldsymbol{R} : x < b\} \tag{4.5}$$

$+\infty$ and $-\infty$ are accumulation points when:

$$+\infty \in D(A) \qquad \Leftrightarrow \forall a \in \boldsymbol{R}, \qquad]a,+\infty[\cap A \text{ is infinite} \tag{4.6}$$

$$-\infty \in D(A) \qquad \Leftrightarrow \forall b \in \boldsymbol{R}, \qquad]-\infty, b[\cap A \text{ is infinite} \tag{4.7}$$

That is, when in every neighborhood of $+\infty$ or $-\infty$ there are infinite points of A (Figure 4.5).

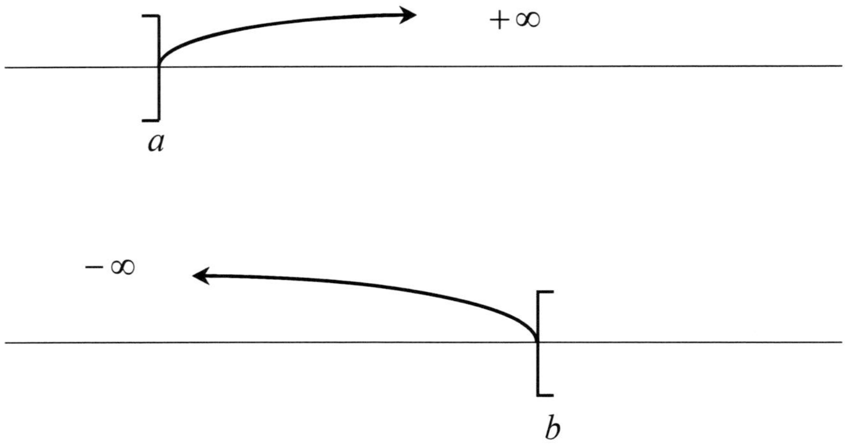

Figure 4.5. Graphical representation of intervals as neighborhoods of $+\infty$ or $-\infty$ as accumulation points.

4.3. Definition of Limit

We define the limit of a function by means of the notions of neighborhood and accumulation points.

Let:

$$f : A \to \boldsymbol{R} \qquad , A \subseteq \boldsymbol{R}$$

$x_0 \in D(A)$ (x_0 accumulation point of A)

We define:

$$\lim_{x \to x_0} f(x) = \lambda \qquad , \lambda \in \boldsymbol{R} \qquad \Leftrightarrow$$

$$\forall U = S(\lambda, \varepsilon), \exists V = S(x_0, \delta) : x \in V \cap A - \{x_0\} \Rightarrow f(x) \in U \tag{4.8}$$

For every neighborhood U of □ there exists a neighborhood V of x_0 such that, if a point x belongs to the intersection of V with A (without the point x_0), then $f(x)$ belongs to U.

In the case of $A \subseteq \boldsymbol{R}$, the definition of limit can be seen graphically in Figure 4.6.

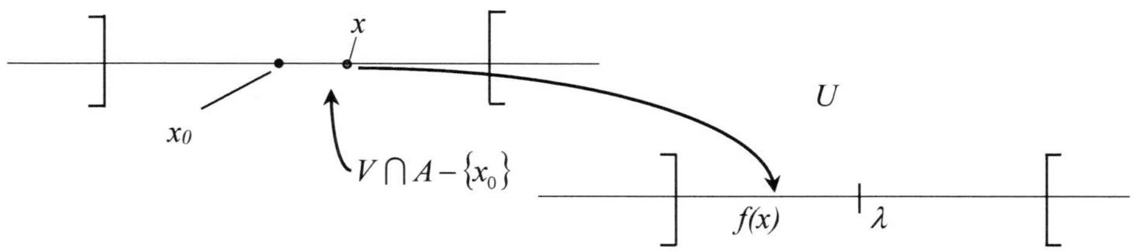

Figure 4.6. Graphical representation of definition of the limit of a function.

Let:

$$f : A \to \boldsymbol{R} \qquad , A \subseteq \boldsymbol{R}$$

$$\lambda \in \boldsymbol{R}, \qquad x_0 \in \boldsymbol{R}, \qquad x_0 \in D(A), \qquad \lim_{x \to x_0} f(x) = \lambda$$

Considering that a point $f(x)$ belongs to the neighborhood U of λ if:

$$U =]\lambda - \varepsilon, \lambda + \varepsilon[$$

$$f(x) \in U \qquad \Leftrightarrow \qquad |f(x) - \lambda| < \varepsilon$$

and similarly, a point x belongs to the neighborhood V of x_0 if:

$$V =]x_0 - \delta, x_0 + \delta[$$

$$x \in V \qquad \Leftrightarrow \qquad |x - x_0| < \delta$$

The definition of limit (Eq. 4.8) becomes:

$$\forall \varepsilon > 0, \exists \delta(\varepsilon) > 0 : x \in A - \{x_0\}, |x - x_0| < \delta(\varepsilon) \Rightarrow |f(x) - \lambda| < \varepsilon \tag{4.9}$$

Which means: for every $\varepsilon > 0$, there exists a $\delta > 0$ (δ depends on ε) such that, if a point x belongs to the set A (excluding point x_0) and the absolute value of the difference between x and x_0 is less than δ, then the absolute value of the difference between $f(x)$ and λ is lower than ε.

4.3.1. Continuous Functions

Let:

$$A \subseteq \boldsymbol{R}, \qquad f : A \rightarrow \boldsymbol{R}, \qquad x_0 \in A,\ x_0 \in D(A)$$

$$f \text{ is } \textit{continuous} \text{ in } x_0 \Leftrightarrow \exists \lim_{x \to x_0} f(x) = f(x_0) \tag{4.10}$$

The function f is continuous in A (and is indicated by $f \in C_A$) if f is continuous in every point of A.

4.3.2. Definition of Limit for Particular Values

Considering the definition of limit (Eq. 4.9):

$$f : A \rightarrow \boldsymbol{R}, \qquad A \subseteq \boldsymbol{R}, \qquad x_0 \in D(A)$$
$$\lim_{x \to x_0} f(x) = \lambda \quad , \lambda \in \boldsymbol{R} \quad \Leftrightarrow$$
$$\forall \varepsilon > 0, \exists \delta(\varepsilon) > 0 : x \in A - \{x_0\}, |x - x_0| < \delta(\varepsilon) \Rightarrow |f(x) - \lambda| < \varepsilon$$

If x_0 or λ take values $+\infty$ or $-\infty$, the definition of limit cannot take the previous form in which they appear $|x - x_0| < \delta(\varepsilon)$ and $|f(x) - \lambda| < \varepsilon$.

In the case of limits that assume values of λ equal to $+\infty$ or $-\infty$, the definition of limit takes the form:

For $\lambda = +\infty$:

$$\forall M \in \boldsymbol{R}, \exists \delta(M) > 0 : x \in A - \{x_0\}, |x - x_0| < \delta(M) \Rightarrow f(x) > M \tag{4.11}$$

For $\lambda = -\infty$:

$$\forall M \in \boldsymbol{R}, \exists \delta(M) > 0 : x \in A - \{x_0\}, |x - x_0| < \delta(M) \Rightarrow f(x) < M \tag{4.12}$$

In the case of limits that assume values of x_0 equal to $+\infty$ or $-\infty$, the definition of limit takes the form:

For $x_0 = +\infty, x \to +\infty$:

$$\forall \varepsilon > 0, \exists M(\varepsilon) \in \boldsymbol{R} : x \in A - \{x_0\}, x > M(\varepsilon) \Rightarrow |f(x) - \lambda| < \varepsilon \tag{4.13}$$

For $x_0 = -\infty$, $x \to -\infty$:

$$\forall \varepsilon > 0, \exists M(\varepsilon) \in \boldsymbol{R} : x \in A - \{x_0\}, x < M(\varepsilon) \Rightarrow |f(x) - \lambda| < \varepsilon \tag{4.14}$$

In general, given the function:

$$f : A \to \boldsymbol{R}, \qquad A \subseteq \boldsymbol{R}, \qquad x_0 \in D(A)$$

there are two equivalent forms for writing the limit of a function:

$$\lim_{x \to x_0} f(x) = \lambda$$

or:

$$f(x) \xrightarrow[x \to x_0]{} \lambda \tag{4.15}$$

4.4. Calculus of Function Limits

4.4.1. Application Examples

Let:

$$f : \boldsymbol{R}^+ \to \boldsymbol{R}, \qquad f(x) = \frac{1}{x},$$
$$\boldsymbol{R}^+ = \{x \in \boldsymbol{R}; x > 0\} =]0, +\infty[$$

The function is a part of hyperbola and is shown in Figure 4.7.

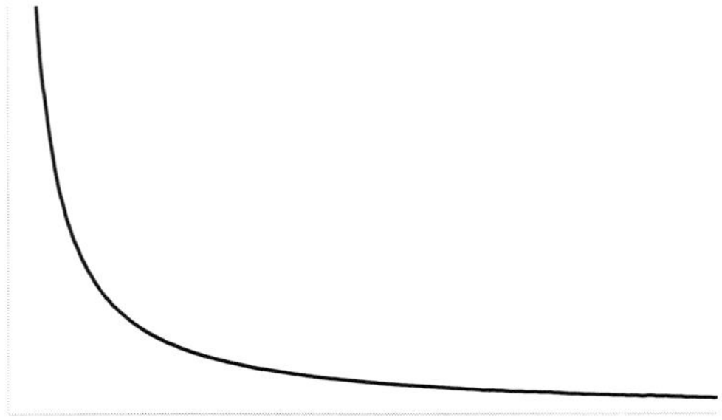

Figure 4.7. Graph of a hyperbola for $x > 0$.

We assume it is worth the limit:

$$\lim_{x \to 0} f(x) = +\infty$$

For it to make sense, it must be:

$$0 \in D(\boldsymbol{R}^+)$$

We consider a neighbourhood of point 0 and arbitrary radius δ (Figure 4.8).

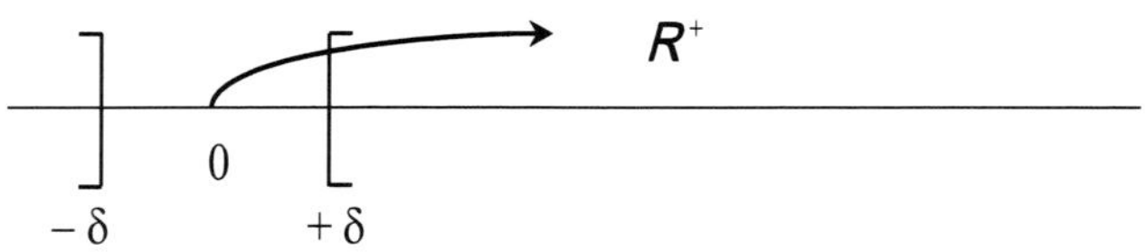

Figure 4.8. Graphical scheme of a neighborhood of point 0 and radius δ.

We have:

$$]-\delta,+\delta[\, \cap \boldsymbol{R}^+ = \,]0,\delta[$$ the interval is infinite

Therefore:

$0 \in D(\boldsymbol{R}^+)$ 0 is an accumulation point

The definition of limit in the case of:

$$f \xrightarrow[x \to 0]{} +\infty$$

is (Eq. 4.11):

$$\forall M \in \boldsymbol{R}, \exists \delta(M) > 0 : x \in A - \{x_0\}, |x - x_0| < \delta(M) \Rightarrow f(x) > M$$

The function is defined for $\boldsymbol{R}^+$ and $x_0 = 0$, so in this case the definition of limit takes the form:

$$\forall M > 0, \exists \delta(M) > 0 : x \in \boldsymbol{R}^+, |x| < \delta(M) \Rightarrow f(x) > M$$

We have:

$$f(x) > M \qquad \Leftrightarrow \qquad \frac{1}{x} > M \qquad \Leftrightarrow \qquad x < \frac{1}{M}$$

Considering that:

$|x| = x$ (the function is defined for $x > 0$)

and assuming:

$$\delta(M) = \frac{1}{M}$$

we have:

$$x < \frac{1}{M} \quad \Rightarrow \quad x < \delta(M) \quad \Rightarrow \quad |x| < \delta(M)$$

The limit exists and is $+\infty$.

Now we assume it is worth the limit:

$$\lim_{x \to +\infty} f(x) = 0$$

The definition of limit in the case of:

$$f \xrightarrow[x \to +\infty]{} 0$$

is (Eq. 4.13):

$$\forall \varepsilon > 0, \exists M(\varepsilon) \in \boldsymbol{R} : x \in A - \{x_0\}, x > M(\varepsilon) \Rightarrow |f(x) - \lambda| < \varepsilon$$

The function is defined for $\boldsymbol{R}^+$ and the limit is $\lambda = 0$, so in this case the definition of limit takes the form:

$$\forall \varepsilon > 0, \exists M(\varepsilon) \in \boldsymbol{R}^+ : x \in \boldsymbol{R}^+, x > M(\varepsilon) \Rightarrow |f(x)| < \varepsilon$$

Considering that:

$|f(x)| = f(x)$ (the function is defined for $x > 0$)

we have:

$$f(x) < \varepsilon \quad \Leftrightarrow \quad \frac{1}{x} < \varepsilon \quad \Leftrightarrow \quad x > \frac{1}{\varepsilon}$$

Assuming:

$$M(\varepsilon) = \frac{1}{\varepsilon}$$

we have:

$$x > \frac{1}{\varepsilon} \quad \Rightarrow \quad x > M(\varepsilon)$$

The limit exists and is $\lambda = 0$.

4.4.2. Methodological Considerations

In general, if we knew or assumed that we knew the value of the limit of a function, it would be necessary to demonstrate its correctness on the basis of the limit definition as it has been done previously for some cases.

By way of example, given the function:

$$f: \boldsymbol{R} \to \boldsymbol{R}, \qquad f(x) = x, \qquad x_0 \in \boldsymbol{R},\ x_0 \in D(\boldsymbol{R})$$

the function is defined on all of ***R***, for which every point $x_0 \in \boldsymbol{R}$ is an accumulation point.

A generic limit for $x \to x_0$ holds:

$$\lim_{x \to x_0} f(x) = \lim_{x \to x_0} x = x_0$$

It should be demonstrated by the definition of limit (Eq. 4.9):

$$\forall \varepsilon > 0, \exists \delta(\varepsilon) > 0 : x \in A - \{x_0\}, |x - x_0| < \delta(\varepsilon) \Rightarrow |f(x) - \lambda| < \varepsilon$$

In this case, we have:

$$f(x) = x, \qquad \lambda = x_0 \qquad \Rightarrow |f(x) - \lambda| = |x - x_0| < \varepsilon$$

So we can assume:

$$\delta(\varepsilon) = \varepsilon$$

The limit exists and is = x_0 and the function *f* is continuous in x_0 (it is continuous on all ***R***).

Normally, one does not know the value of a limit whose correctness can be demonstrated on the basis of the definition of limit as it has been done previously.

Operationally, the limit can be calculated by working on the expression of the function and trying to express it in terms of simpler functions whose limit is already known.

In this sense, some limits will be considered known and these limits can be used in case of more complex functions.

For example:

$$f: \boldsymbol{R} \to \boldsymbol{R}, \qquad f(x) = x^n, \qquad n \in \boldsymbol{N}, \qquad x_0 \in \boldsymbol{R}$$

$$\lim_{x \to x_0} f(x) = \lim_{x \to x_0} x^n = x_0^{\ n}$$

x_0 is an accumulation point and ed *f* is continuous in x_0 (it is continuous on all ***R***).

4.5. Properties of Limits

The limit of a function for $x \to x_0$, if it exists, is unique:

$$f: A \to \boldsymbol{R}, \qquad A \subseteq \boldsymbol{R}, \qquad x_0 \in D(A)$$
$$\lim_{x \to x_0} f(x) = \lambda, \qquad \lim_{x \to x_0} f(x) = \mu \qquad \Rightarrow \lambda = \mu$$
$$\lim_{x \to x_0} f(x) = \lambda \quad \equiv \quad f(x) \xrightarrow[x \to x_0]{} \lambda$$

Let:

$$A \subseteq \boldsymbol{R}, \qquad x_0 \in D(A), \qquad f, g: A \to \boldsymbol{R}$$
$$f(x) \xrightarrow[x \to x_0]{} \alpha, \qquad g(x) \xrightarrow[x \to x_0]{} \beta, \qquad \alpha, \beta \in \boldsymbol{R}$$

The following properties apply:

$$f(x) + g(x) \xrightarrow[x \to x_0]{} \alpha + \beta \tag{4.16a}$$

$$f(x) \cdot g(x) \xrightarrow[x \to x_0]{} \alpha \cdot \beta \tag{4.16b}$$

$$f(x) \neq 0, \forall x \in A, \alpha \neq 0 \qquad \Rightarrow \frac{1}{f(x)} \xrightarrow[x \to x_0]{} \frac{1}{\alpha} \tag{4.16c}$$

$$f(x) \leq g(x), \forall x \in A \qquad \Rightarrow \alpha \leq \beta \tag{4.16d}$$

$$\begin{gathered} h: A \to \boldsymbol{R}, f(x) \leq h(x) \leq g(x), \forall x \in A, \alpha = \beta \\ \Rightarrow h(x) \xrightarrow[x \to x_0]{} \alpha = \beta \end{gathered} \tag{4.16e}$$

$$|f(x)| \xrightarrow[x \to x_0]{} |\alpha| \tag{4.16f}$$

4.6. LIMIT FOR *X* THAT TENDS TO X_0 FROM THE RIGHT OR FROM THE LEFT

Let:

$$f : A \to \boldsymbol{R}, \quad A \subseteq \boldsymbol{R}, x_0 \in D(A)$$

We define limit for x that tends to x_0 from the right or right limit:

$$\lim_{x \to x_0^+} f(x) \equiv \lim_{\substack{x \to x_0 \\ x > x_0}} f(x) \tag{4.17}$$

We define limit for x that tends to x_0 from the left or left limit:

$$\lim_{x \to x_0^-} f(x) \equiv \lim_{\substack{x \to x_0 \\ x < x_0}} f(x) \tag{4.18}$$

It exists the limit of the function if and only if the limit from the right is equal to the limit from the left:

$$\lim_{x \to x_0^+} f(x) = \lim_{x \to x_0^-} f(x) = \lambda \quad \Leftrightarrow \quad \lim_{x \to x_0} f(x) = \lambda \tag{4.19}$$

4.7. APPLICATION EXAMPLES OF LIMIT FROM RIGHT AND LEFT

4.7.1. Example 1

Let:

$$f : \boldsymbol{R} - \{0\} \to \boldsymbol{R}, \qquad f(x) = \frac{1}{x}$$

The definition domain of the function consists of the function domain in the numerator, in this case $\boldsymbol{R}$, minus the values in which the denominator is equal to zero: $x = 0$.

The function is then defined by:

$$f : \boldsymbol{R} - \{0\} \rightarrow \boldsymbol{R}$$

An alternative writing form of domain can be:

$$f :]-\infty, 0[\cup]0, +\infty[\rightarrow \boldsymbol{R}$$

The graph of the function is shown in Figure 4.9:

We calculate the limits for $x \rightarrow 0^+$ and $x \rightarrow 0^-$.

$\lim\limits_{x \rightarrow 0^+} \dfrac{1}{x} = +\infty$ (as seen above, subsection 4.4.1)

For $x \rightarrow 0^-$ we could calculate the limit in a similar way (using the definition of limit, Eq. 4.10 – 4.14), or we can make some considerations using one of the properties of the limits (Eq. 4.16b):

$$f(x) \cdot g(x) \xrightarrow[x \rightarrow x_0]{} \alpha \cdot \beta$$

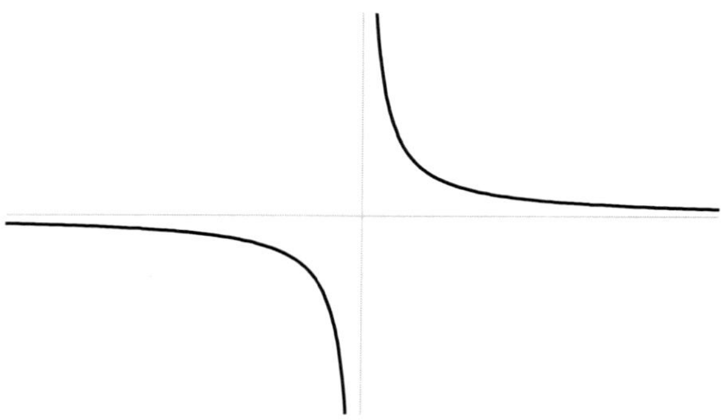

Figure 4.9. Graph of a hyperbola.

The limit of the product is equal to the product of the limits.

This is true even considering the function $g(x)$ as a constant a:

$$\lim_{x \rightarrow 0^-} \left(a \cdot \frac{1}{x} \right) = a \cdot \lim_{x \rightarrow 0^-} \frac{1}{x}$$

the limit of the product of a constant for a function *f(x)* is equal to the product of the constant for the limit of the function.

By considering $a = -1$ we have:

$$\frac{1}{x}=(-1)\cdot(-1)\cdot\frac{1}{x}$$

in particular:

$$(-1)\cdot\frac{1}{x}=-\frac{1}{x}\quad(x<0)\quad\rightarrow\quad=\frac{1}{x}\quad(x>0)$$

the limit becomes for $x > 0$, so multiplying the limit of the function by – 1, we obtain:

$$-1\cdot \lim_{x\to 0^-}\frac{1}{x}=\lim_{x\to 0^-}\left(-1\cdot\frac{1}{x}\right)=\lim_{x\to 0^+}\frac{1}{x}=+\infty$$

from which:

$$\lim_{x\to 0^-}\frac{1}{x}=\lim_{x\to 0^-}\left[(-1)\cdot(-1)\cdot\frac{1}{x}\right]=(-1)\cdot\lim_{x\to 0^-}\left[(-1)\cdot\frac{1}{x}\right]=$$
$$=(-1)\cdot(+\infty)=-\infty$$

In this case:

$$\lim_{x\to 0^+}\frac{1}{x}\neq\lim_{x\to 0^-}\frac{1}{x}$$

thus:

$\lim_{x\to 0}\frac{1}{x}$ it does not exist

4.7.2. Example 2

Let:

$$f:\boldsymbol{R}-\{0\}\rightarrow\boldsymbol{R},\qquad f(x)=\frac{1}{|x|},$$

The graph of the function is shown in Figure 4.10:
We want to calculate:

$$\lim_{x \to 0} \frac{1}{|x|}$$

The absolute value function is defined as (Eq. 3.16):

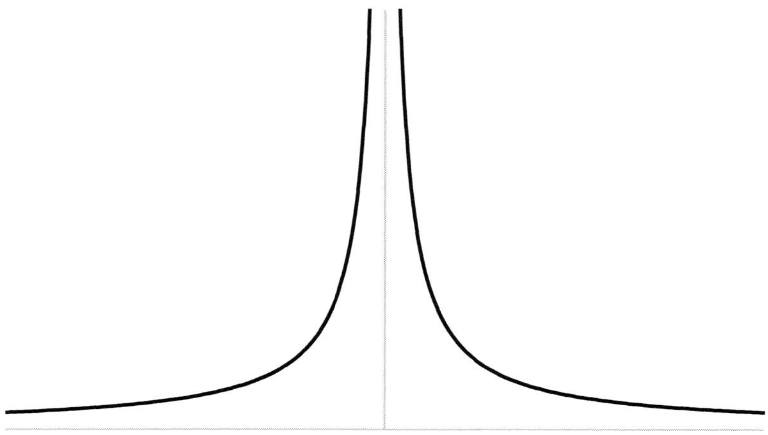

Figure 4.10. Graph of the absolute value of a hyperbola.

$$|x| = \begin{matrix} x & ,x \geq 0 \\ -x & ,x \leq 0 \end{matrix}$$

The limit is split out "from the right" and "from the left":

$$\lim_{x \to 0^+} \frac{1}{|x|} = +\infty \text{ (as already seen)}$$

$$\lim_{x \to 0^-} \frac{1}{|x|} = \lim_{x \to 0^-} \frac{1}{-x} = \lim_{x \to 0^-} \left(-1 \cdot \frac{1}{x} \right) = \lim_{x \to 0^+} \frac{1}{x} = +\infty$$

Thus:

$$\lim_{x \to 0^+} \frac{1}{|x|} = \lim_{x \to 0^-} \frac{1}{|x|} = \lim_{x \to 0} \frac{1}{|x|} = +\infty$$

The limit exists.

4.8. Exponential Function

Let:

$x \in \boldsymbol{R}$, $a \in \boldsymbol{R}$, $a =$ constant

We want to define the function:

$$f(x) = a^x \tag{4.20}$$

If x is a real number, it cannot be $a < 0$. For example, a square root cannot be calculated for a negative number:

$$a^{\frac{1}{2}} = \sqrt{a}$$

Let's assume:

$$a \in \boldsymbol{R}^+$$

$a = 0$ is not considered, because $0^x = 0$.
Let's assume in particular::

$$x \in \boldsymbol{R}, \qquad f(x) = a^x, \qquad a > 1 \tag{4.21}$$

The following properties apply:

$$\begin{aligned} &a^{x+y} = a^x a^y \\ &\left(a^x\right)^y = a^{xy} \\ &x < y \Rightarrow a^x < a^y \ (a > 1) \end{aligned} \tag{4.22}$$

Properties are worth if $x \in \boldsymbol{N}$:

$a^x = a \cdot a \cdot a \cdots\cdots a$ (a multiplied by itself x times)

If:

$x \in \boldsymbol{I}$

The previous properties apply. Furthermore:

$$a^x = \begin{cases} a^x & ,x \in \boldsymbol{N} \\ 1 & ,x = 0 \\ \dfrac{1}{a^{|x|}} & ,x \in -\boldsymbol{N} \end{cases} \tag{4.23}$$

In general:

$$a^x = \frac{1}{a^{-x}} \tag{4.24}$$

If:

$$x \in \boldsymbol{Q},\ \boldsymbol{Q} = \left\{ \frac{m}{n} : m \in \boldsymbol{I}, n \in \boldsymbol{N} \right\}$$

The previous properties apply. Furthermore:

$$x = \frac{m}{n} \qquad \rightarrow \qquad a^x = a^{\frac{m}{n}} = \left(a^{\frac{1}{n}} \right)^m$$

From the definition of n-th root:

$$y = \sqrt[n]{a} \qquad \Leftrightarrow \qquad y^n = a$$

$$\left(\sqrt[n]{a} \right)^n = a$$

Let's say by definition:

$$\sqrt[n]{a} = a^{\frac{1}{n}} \qquad \rightarrow \qquad \left(\sqrt[n]{a} \right)^n = \left(a^{\frac{1}{n}} \right)^n = a^{\frac{n}{n}} = a \tag{4.25}$$

Raising a number to $\frac{1}{n}$ it is equivalent to calculating its n-th root.

If:

$x \in \boldsymbol{R}$

All previous properties apply.

Let:

$$f : \boldsymbol{R} \to \boldsymbol{R}, \qquad f(x) = a^x, \qquad x \in \boldsymbol{R}, a \in \boldsymbol{R}^+, a > 1$$

The following limits apply:

$$\lim_{x \to 0} a^x = a^0 = 1 \tag{4.26}$$

$$\lim_{x \to +\infty} a^x = +\infty \tag{4.27}$$

$$\lim_{x \to -\infty} a^x = \lim_{x \to -\infty} a^{-|x|} = \lim_{x \to -\infty} \frac{1}{a^{|x|}} = 0 \tag{4.28}$$

The graph of the function is shown in Figure 4.11:

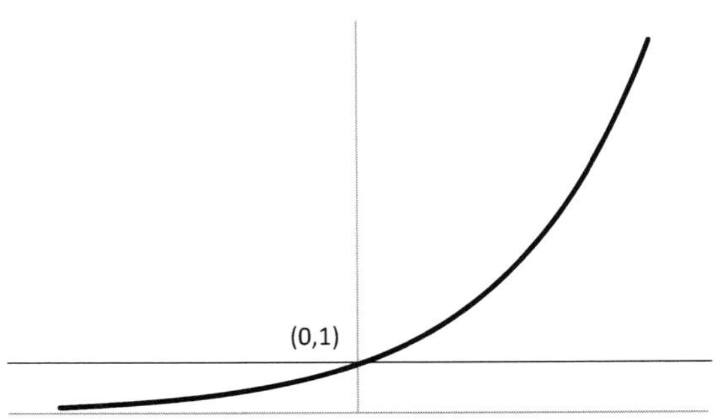

Figure 4.11. Graph of the exponential function: $a^x, a > 1$.

Let:

$$f : \boldsymbol{R} \to \boldsymbol{R}, \qquad f(x) = a^x, \qquad x \in \boldsymbol{R}, a \in \boldsymbol{R}^+, a < 1$$

The limit applies:

$$\lim_{x \to 0} a^x = a^0 = 1 \tag{4.29}$$

For the limits for $x \to +\infty$ and $x \to -\infty$, we can consider that:

$$a = \frac{1}{a^{-1}}$$

and:

$$a^x = \left(\frac{1}{a^{-1}}\right)^x = \frac{1}{\left(a^{-1}\right)^x}$$

If:

$a < 1 \Rightarrow a^{-1} > 1$ (if a number is <1, its reciprocal is >1)

For $x \to +\infty$ it can be observed that:

$$\lim_{x \to +\infty} \left(a^{-1}\right)^x = +\infty \quad (a^{-1} > 1)$$

thus:

$$\lim_{x \to +\infty} a^x = \lim_{x \to +\infty} \frac{1}{\left(a^{-1}\right)^x} = \frac{1}{+\infty} = 0 \tag{4.30}$$

For $x \to -\infty$ it can be observed that:

$$\lim_{x \to -\infty} \left(a^{-1}\right)^x = 0 \quad (a^{-1} > 1)$$

thus:

$$\lim_{x \to -\infty} a^x = \lim_{x \to -\infty} \frac{1}{\left(a^{-1}\right)^x} = \frac{1}{0} = +\infty \tag{4.31}$$

The graph of the function is shown in Figure 4.12:

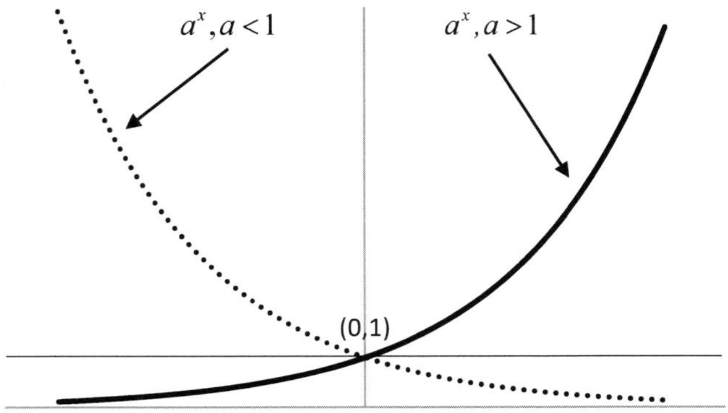

Figure 4.12. Graph of the exponential function: a^x, a >1 and a <1.

4.9. Logarithm Function

Let:

$x \in \boldsymbol{R}^+$, $a \in \boldsymbol{R}^+$, $a =$ constant

We define the *logarithm of x to the base a*:

$f : \boldsymbol{R}^+ \to \boldsymbol{R} \qquad , a \in \boldsymbol{R}^+$

$$y = f(x) = \log_a x \qquad \Leftrightarrow \qquad a^y = x \tag{4.32}$$

That is, y is the exponent to be given to the base a to obtain x.

The exponential function and the logarithm function are one the inverse function of the other. For this statement to be true, the exponential function should be more correctly defined as:

$f : \boldsymbol{R} \to \boldsymbol{R}^+, \qquad f(x) = a^x, \qquad x \in \boldsymbol{R}, a \in \boldsymbol{R}^+$

(and not $f : \boldsymbol{R} \to \boldsymbol{R}$)

We should consider the two cases $a > 1$ and $a < 1$, but we only consider the case $a > 1$, because it is the one that occurs most frequently.

From the definition of logarithm function (Eq. 4.32) we have:

for $x = 1$:

$$y = \log_a 1 = 0 \quad \left(a^y = 1 \quad \Leftrightarrow \quad y = 0 \Rightarrow a^0 = 1\right) \tag{4.33}$$

for $x \to +\infty$:

$$\lim_{x \to +\infty} \log_a x = +\infty \left(\lim_{y \to +\infty} a^y = +\infty \right) \tag{4.34}$$

for $x \to 0^+$:

$$\lim_{x \to 0^+} \log_a x = -\infty \left(\lim_{y \to -\infty} a^y = 0 \right) \tag{4.35}$$

The limit for $x \to 0^+$ is due to the fact that $x \in \boldsymbol{R}^+$ (the logarithm function is not defined for negative values of *x*).

The graph of the logarithm function is shown in Figure 4.13:

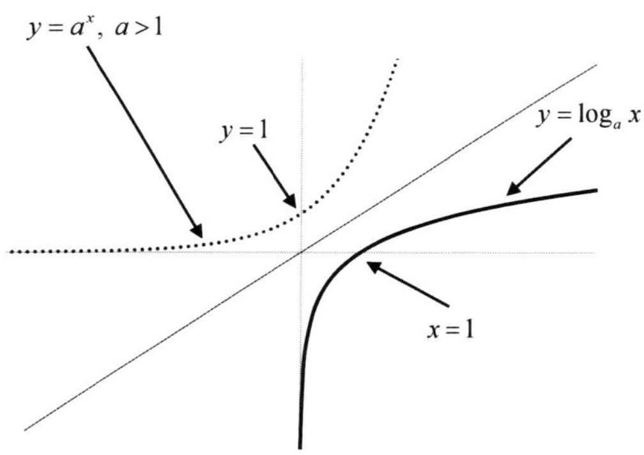

Figure 4.13. Graph of the logarithm function: $\log_a x$.

4.10. Properties of the Logarithm Function

The following properties apply to the logarithm function.

$$f : \boldsymbol{R}^+ \to \boldsymbol{R} \qquad , b, c, a \in \boldsymbol{R}^+$$

$$\log_a (b \cdot c) = \log_a b + \log_a c \quad b, c \in \boldsymbol{R}^+ \tag{4.36}$$

$$\log_a b^\alpha = \alpha \log_a b \quad b \in \boldsymbol{R}^+, \alpha \in \boldsymbol{R} \tag{4.37}$$

To verify the first property, both members of the Equation 4.36 are considered as the argument of the exponential function in the base a:

$$a^{\lg_a(b\cdot c)} = a^{\lg_a b+\lg_a c} = a^{\lg_a b}\cdot a^{\lg_a c}$$

The exponential function is the inverse function of the logarithm, so:

$(b\cdot c)=(b)\cdot(c)$ the first property is verified.

As before, to verify the second property both members of the Equation 4.37 are considered as the argument of the exponential function in the base a:

$$a^{\log_a b^{\alpha}} = a^{\alpha\log_a b} = \left(a^{\log_a b}\right)^{\alpha}$$

Thus:

$b^{\alpha} = (b)^{\alpha}$ the second property is verified.

4.11. Natural Logarithms and Logarithms to the Base 10

4.11.1. Napier's Number and Logarithms Definition

Napier's number "e" is defined as the limit of the sequence:

$$(a_n)_{n\in N} = \left[\left(1+\frac{1}{n}\right)^n\right]_{n\in N} \tag{4.38a}$$

$$\lim_{n\to+\infty}\left(1+\frac{1}{n}\right)^n = e \tag{4.38b}$$

Or as the limit of the real variable function:

$$\lim_{z\to+\infty}\left(1+\frac{1}{z}\right)^z = e \qquad ,z\in \boldsymbol{R} \tag{4.39a}$$

$$\lim_{z \to -\infty}\left(1+\frac{1}{z}\right)^{z} = e \qquad , z \in \boldsymbol{R} \tag{4.39b}$$

where, by approximating the real number:

$$e = 2.718281 \tag{4.40}$$

The *natural logarithm*, or simply *logarithm*, is defined as the function (the base is constituted by the Napier's number):

$$y = f(x) = \lg x \equiv \log x \equiv \ln x \quad \Leftrightarrow \quad e^{y} = x \tag{4.41}$$

The *logarithm to base* 10, or *decimal logarithm*, or *Briggsian logarithm*, or simply *common logarithm*, is defined as the function (the base is constituted by the number 10):

$$y = f(x) = \lg_{10} x \equiv \log_{10} x \equiv Logx \quad \Leftrightarrow \quad 10^{y} = x \tag{4.42}$$

4.11.2. Conversion of Natural Logarithm to Common Logarithm

Let be given the definition of the logarithm of x to the base a (Eq. 4.32), we assume:

$$e^{y} = x \tag{4.43}$$

$$y = \log x \tag{4.44}$$

Both members of Eq. 4.43 are considered as the argument of the log_{10} function:

$$\log_{10} e^{y} = \log_{10} x$$

From which (Eq. 4.44):

$$\log_{10} e^{y} = y \cdot \log_{10} e = \log x \cdot \log_{10} e$$

Thus:

$$\log_{10} x = \left(\log_{10} e\right) \cdot \log x \tag{4.45}$$

where, by approximating the real number, the conversion factor is:

$$\log_{10} e = 0.43429$$

The opposite conversion is given by:

$$\log x = (\log_{10} e)^{-1} \cdot \log_{10} x \qquad (4.46)$$

where, by approximating the real number, the conversion factor is:

$$(\lg_{10} e)^{-1} = 2.30259$$

4.12. Power Function

We have seen previously (subsection 4.4.2) that, given the function:

$$f : \boldsymbol{R} \to \boldsymbol{R}, \qquad f(x) = x^n, \qquad n \in \boldsymbol{N}, \qquad x_0 \in \boldsymbol{R}$$

$$\lim_{x \to x_0} f(x) = \lim_{x \to x_0} x^n = x_0^{\,n}$$

x_0 is an accumulation point and f is continuous in x_0 (f is continuous on $\boldsymbol{R}$).

We define the power function considering the base as a variable and the exponent as a real constant:

$$f : \boldsymbol{R}^+ \to \boldsymbol{R}, \qquad f(x) = x^a, \qquad a \in \boldsymbol{R} \qquad (4.47)$$

The domain of definition is $\boldsymbol{R}^+$, since being the exponent a real number, cases such as for example an even root of a negative number cannot be defined.

Furthermore, if we considered $x = 0$ and a = 0, the function would be included in case 0^0.

4.13. Trigonometric Functions

From the definition of function, the conditions for f to be a function are (Eq. 3.1):

$$f \subseteq A \times B$$

$$\forall a \in A, \exists! b \in B : b = f(a)$$

In the case of a circumference represented in a Cartesian coordinate system (Figure 4.14):

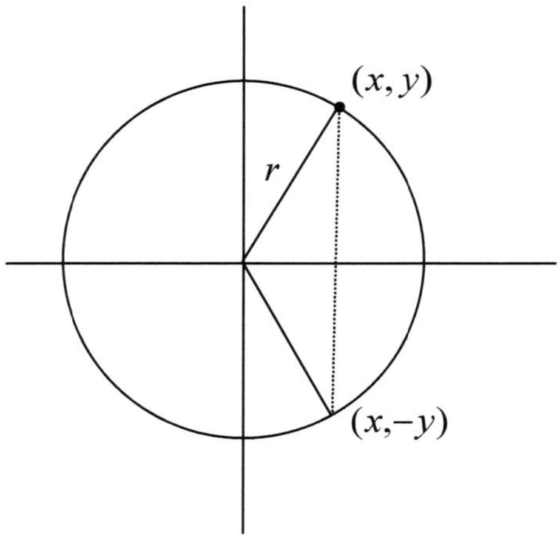

Figure 4.14. Representation of two coordinates (x, y) on a circumference.

The equation that represents a generic point on a circumference cannot be seen as a function:

$$r = \sqrt{x^2 + y^2} \quad (r = \text{circumference radius})$$

In fact, the equation:

$$r^2 = x^2 + y^2 \quad \rightarrow \quad y^2 = r^2 - x^2$$

it admits two solutions:

$$y = \pm\sqrt{r^2 - x^2}$$

However, if one considers to identify a point on the circumference not on the basis of its Cartesian coordinates but on the basis of the radius of the circumference and of the angle that represents the width of the arc corresponding to the point being considered, giving the circumference a counterclockwise orientation (Figure 4.15), each value of an angle corresponds to a single point on the circumference.

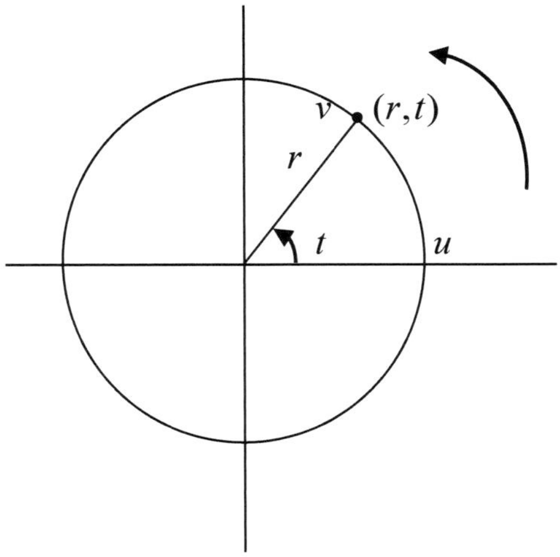

Figure 4.15. Representation of a point on a circumference.

Different systems of units have been defined for the measurement of an angle.

In particular, we define the angle measured in radians (counterclockwise) as the length of an arc on a circumference of unit radius:

$$t = l(\widehat{uv}) \tag{4.48}$$

The arc length corresponding to the unit circle is 2π, where, by approximating the real number, $\pi = 3.1415927$.

An angle measured in radians belongs interval:

$$t \in [0, 2\pi[\tag{4.49}$$

Given a generic real number, it holds:

$$t \in \boldsymbol{R} \quad \Rightarrow \quad \exists! k \in \boldsymbol{I}, \tau \in [0, 2\pi[: t = \tau + 2k\pi \tag{4.50}$$

4.14. Sine and Cosine Functions

We consider the unit circle (Figure 4.16):

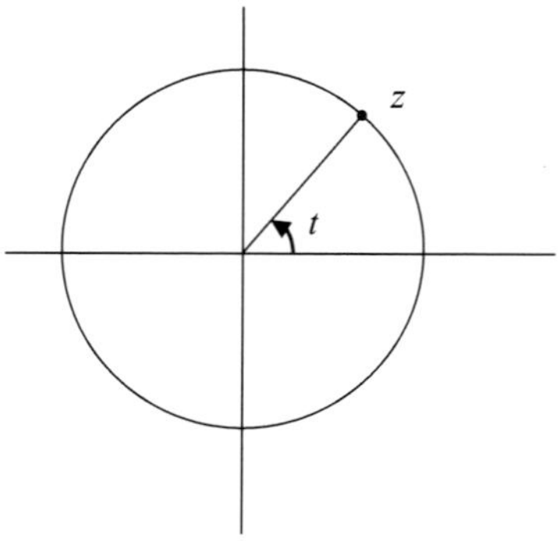

Figure 4.16. Representation of a point on a unit circle.

As seen previously (Eq. 4.50):

$$t \in \boldsymbol{R} \quad \Rightarrow \quad \exists! k \in \boldsymbol{I}, \tau \in [0, 2\pi[: \quad t = \tau + 2k\pi$$

We set by definition:

$$x = \cos t, \qquad y = \sin t \tag{4.51}$$

where *sin* and *cos* indicate respectively the *sine* and *cosine* functions.

We define:

$$\cos t = \cos(\tau + 2k\pi) = \cos\tau \tag{4.52}$$

$$\sin t = \sin(\tau + 2k\pi) = \sin\tau \tag{4.53}$$

$$\sin^2 t + \cos^2 t = 1 \tag{4.54}$$

$$\sin 0 = 0 \quad , \quad \cos 0 = 1 \tag{4.55}$$

$$\sin\frac{\pi}{2} = 1 \quad , \quad \cos\frac{\pi}{2} = 0 \tag{4.56}$$

$$\sin\pi = 0 \quad , \quad \cos\pi = -1 \tag{4.57}$$

$$\sin\frac{3\pi}{2} = -1 \quad , \quad \cos\frac{3\pi}{2} = 0 \tag{4.58}$$

$$\sin 2\pi = 0 \quad , \quad \cos 2\pi = 1 \tag{4.59}$$

Sine and cosine functions are graphically represented in Figures 4.17 and 4.18.

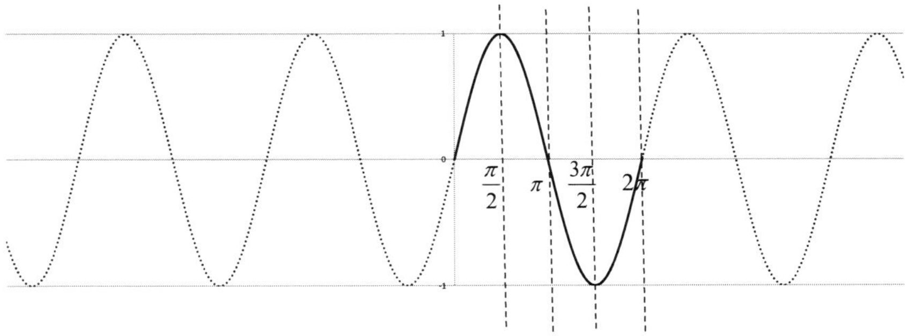

Figure 4.17. Graph of the sine function.

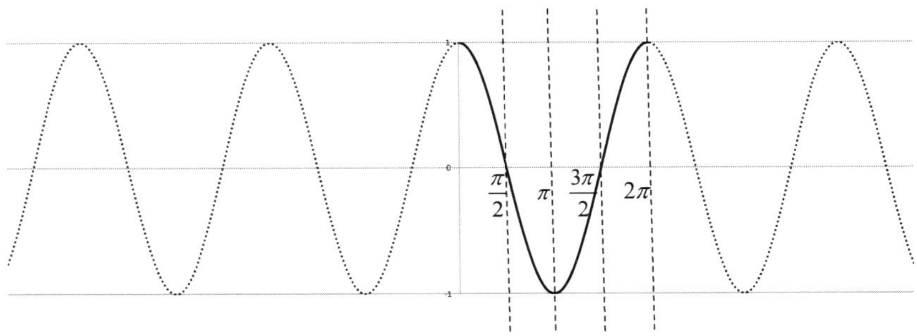

Figure 4.18. Graph of the cosine function.

4.15. Tangent and Cotangent Functions

We define the *tangent* function:

$$\tan t = \frac{\sin t}{\cos t} \tag{4.60}$$

We define the *cotangent* function:

$$\cot t = \frac{\cos t}{\sin t} \tag{4.61}$$

In general, it holds (Eq. 4.55 – 4.59):

$$\lim_{t \to 0} \sin t = \sin 0 = 0$$

$$\lim_{t \to \frac{\pi}{2}} \sin t = \sin \frac{\pi}{2} = 1$$

$$\lim_{t \to \pi} \sin t = \sin \pi = 0$$

$$\lim_{t \to \frac{3\pi}{2}} \sin t = \sin \frac{3\pi}{2} = -1$$

$$\lim_{t \to 2\pi} \sin t = \sin 2\pi = 0$$

$$\lim_{t \to 0} \cos t = \cos 0 = 1$$

$$\lim_{t \to \frac{\pi}{2}} \cos t = \cos \frac{\pi}{2} = 0$$

$$\lim_{t \to \pi} \cos t = \cos \pi = -1$$

$$\lim_{t \to \frac{3\pi}{2}} \cos t = \cos \frac{3\pi}{2} = 0$$

$$\lim_{t \to 2\pi} \cos t = \cos 2\pi = 1$$

The sine and cosine functions are continuous functions.

If we calculate the previous limits for tangent and cotangent functions we can see that these are not continuous functions.

4.15.1. Tangent Function

For the tangent function we want to calculate the limits:

$$\lim_{t \to \frac{\pi}{2}} \tan t = \lim_{t \to \frac{\pi}{2}} \frac{\sin t}{\cos t}, \quad \lim_{t \to \frac{3\pi}{2}} \tan t = \lim_{t \to \frac{3\pi}{2}} \frac{\sin t}{\cos t}$$

As seen before:

$$\lim_{t \to \frac{\pi}{2}} \cos t = \cos\frac{\pi}{2} = 0, \quad \lim_{t \to \frac{3\pi}{2}} \cos t = \cos\frac{3\pi}{2} = 0$$

In general:

$$\lim_{t \to \left(\frac{\pi}{2} + k\pi\right)} \cos t = \lim_{t \to \left(k + \frac{1}{2}\right)\pi} \cos t = 0, \; k \in \boldsymbol{I} \tag{4.62}$$

So, since the cosine function is at the denominator, the tangent function has discontinuities for:

$$t \to \left(k + \frac{1}{2}\right)\pi \quad , \quad k \in \boldsymbol{I} \tag{4.62a}$$

It is necessary to verify if the function has a limit at these points, i.e., if:

$$\lim_{t \to \left(k + \frac{1}{2}\right)\pi^+} \tan t = \lim_{t \to \left(k + \frac{1}{2}\right)\pi^-} \tan t = \lim_{t \to \left(k + \frac{1}{2}\right)\pi} \tan t$$

We can consider the values $t = \dfrac{\pi}{2}$ and $t = \dfrac{3}{2}\pi$.

For $t = \dfrac{\pi}{2}$ the sine function at the numerator has limit:

$$\lim_{t \to \frac{\pi}{2}} \sin t = \sin\frac{\pi}{2} = 1$$

The cosine function has limits:

$$\lim_{t \to \frac{\pi}{2}^+} \cos t = 0 \text{, for } t > \frac{\pi}{2}: \ \cos t < 0$$

$$\lim_{t \to \frac{\pi}{2}^-} \cos t = 0 \text{, for } t < \frac{\pi}{2}: \ \cos t > 0$$

Thus (the numerator is positive):

$$\lim_{t \to \frac{\pi}{2}^+} \tan t = \lim_{t \to \frac{\pi}{2}^+} \frac{\sin t}{\cos t} = \frac{1}{0^-} = -\infty \tag{4.63}$$

For the right limit the denominator tends to 0 by negative values.

$$\lim_{t \to \frac{\pi}{2}^-} \tan t = \lim_{t \to \frac{\pi}{2}^-} \frac{\sin t}{\cos t} = \frac{1}{0^+} = +\infty \tag{4.64}$$

For the left limit the denominator tends to 0 by positive values.

For $t = \frac{3\pi}{2}$ the sine function at the numerator has limit:

$$\lim_{t \to \frac{3\pi}{2}} \sin t = \sin \frac{3\pi}{2} = -1$$

The cosine function has limits:

$$\lim_{t \to \frac{3\pi}{2}^+} \cos t = 0 \text{, for } t > \frac{3\pi}{2}: \cos t > 0$$

$$\lim_{t \to \frac{3\pi}{2}^-} \cos t = 0 \text{, for} t < \frac{3\pi}{2}: \ \cos t < 0$$

Thus (the numerator is negative):

$$\lim_{t \to \frac{3\pi}{2}^+} \tan t = \lim_{t \to \frac{3\pi}{2}^+} \frac{\sin t}{\cos t} = \frac{-1}{0^+} = -\infty \tag{4.65}$$

For the right limit the denominator tends to 0 by positive values.

$$\lim_{t \to \frac{3\pi}{2}^-} \tan t = \lim_{t \to \frac{3\pi}{2}^-} \frac{\sin t}{\cos t} = \frac{-1}{0^-} = +\infty \tag{4.66}$$

For the left limit the denominator tends to 0 by negative values.

Thus, for values $t = \frac{\pi}{2}$ and $t = \frac{3}{2}\pi$ the tangent function has no limit.

This behavior repeats periodically for:

$$t \to \left(k + \frac{1}{2}\right)\pi \quad , \quad k \in I$$

Figure 4.19 shows the tangent function.

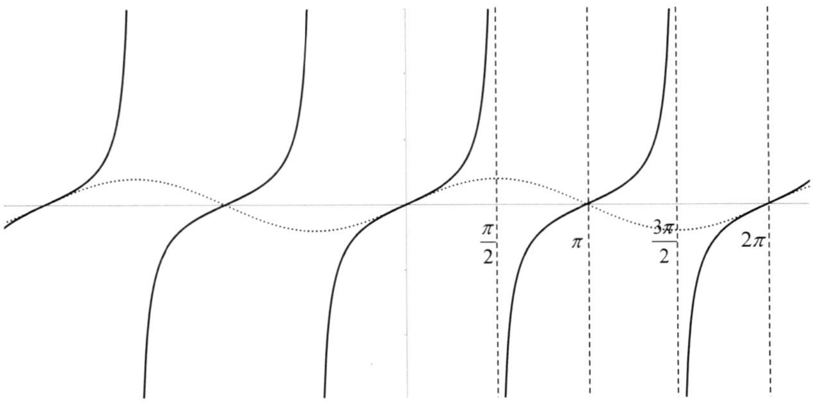

Figure 4.19. Graph of tangent function (in dashed line the sine function).

4.15.2. Cotangent Function

For the cotangent function we want to calculate the limits:

$$\lim_{t \to 0} \cot t = \lim_{t \to 0} \frac{\cos t}{\sin t}, \ \lim_{t \to \pi} \cot t = \lim_{t \to \pi} \frac{\cos t}{\sin t}$$

As seen before:

$$\lim_{t\to 0}\sin t = \sin 0 = 0,\ \lim_{t\to \pi}\sin t = \sin\pi = 0$$

In general:

$$\lim_{t\to k\pi}\sin t = 0,\ k \in I \tag{4.67}$$

So, since the sine function is at the denominator, the cotangent function has discontinuities for:

$$t \to k\pi \quad , \quad k \in I \tag{4.67a}$$

It is necessary to verify if the function has a limit at these points, i.e., if:

$$\lim_{t\to k\pi^+}\cot t = \lim_{t\to k\pi^-}\cot t = \lim_{t\to k\pi}\cot t$$

We can consider the values $t = 0$ and $t = \pi$.

For $t = 0$ the cosine function at the numerator has limit:

$$\lim_{t\to 0}\cos t = \cos 0 = 1$$

The sine function has limits:

$$\lim_{t\to 0^+}\sin t = 0, \text{ for } t > 0:\ \sin t > 0$$

$$\lim_{t\to 0^-}\sin t = 0, \text{ for } t < 0:\ \sin t < 0$$

Thus (the numerator is positive):

$$\lim_{t\to 0^+}\cot t = \lim_{t\to 0^+}\frac{\cos t}{\sin t} = \frac{1}{0^+} = +\infty \tag{4.68}$$

For the right limit the denominator tends to 0 by positive values.

$$\lim_{t\to 0^-} \cot t = \lim_{t\to 0^-} \frac{\cos t}{\sin t} = \frac{1}{0^-} = -\infty \tag{4.69}$$

For the left limit the denominator tends to 0 by negative values.
For $t = \pi$ the cosine function at the numerator has limit:

$$\lim_{t\to\pi} \cos t = \cos 0 = -1$$

The sine function has limits:

$$\lim_{t\to\pi^+} \sin t = 0 \text{, for } t > \pi : \sin t < 0$$

$$\lim_{t\to\pi^-} \sin t = 0 \text{, for } t < \pi : \sin t > 0$$

Thus (the numerator is negative):

$$\lim_{t\to\pi^+} \cot t = \lim_{t\to\pi^+} \frac{\cos t}{\sin t} = \frac{-1}{0^-} = +\infty \tag{4.70}$$

For the right limit the denominator tends to 0 by negative values.

$$\lim_{t\to\pi^-} \cot t = \lim_{t\to\pi^-} \frac{\cos t}{\sin t} = \frac{-1}{0^+} = -\infty \tag{4.71}$$

For the left limit the denominator tends to 0 by positive values.
Thus, for values $t = 0$ and $t = \pi$ the tangent function has no limit.
This behavior repeats periodically for:

$$t \to k\pi \quad , \quad k \in \boldsymbol{I}$$

Figure 4.20 shows the cotangent function.

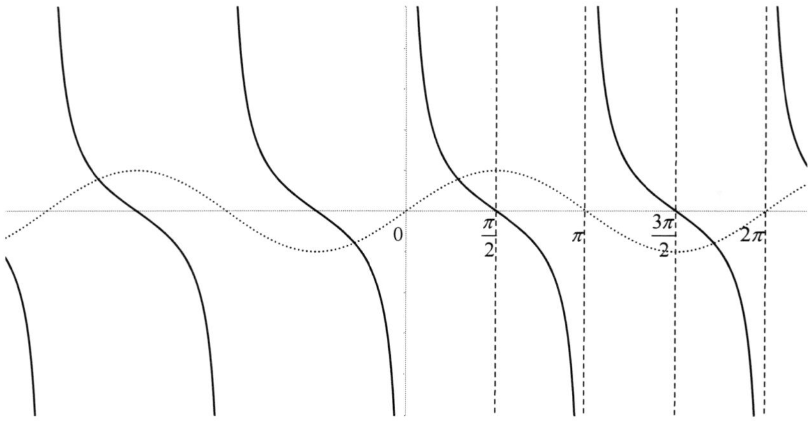

Figure 4.20. Graph of cotangent function (in dashed line the sine function).

Chapter 5

DERIVATIVE OF A FUNCTION

5.1. INCREMENTAL RATIO AND DERIVATIVE OF A FUNCTION AT A POINT

Let:

$$f : A \to \mathbf{R}, \qquad A \subseteq \mathbf{R}, \qquad x_0 \in A \cap D(A)$$

We define the function:

$$\varphi(x) = \frac{f(x) - f(x_0)}{x - x_0}, \qquad x \in A - \{x_0\} \tag{5.1}$$

The function is defined in $A - \{x_0\}$ and it is called the *incremental ratio of f* starting from point x_0.

We define the limit:

$$\lim_{x \to x_0} \varphi(x) = \lim_{x \to x_0} \frac{f(x) - f(x_0)}{x - x_0} \tag{5.2}$$

The geometric meaning of the limit of the incremental ratio is to represent the tangent to the function $f(x)$ at point x_0:

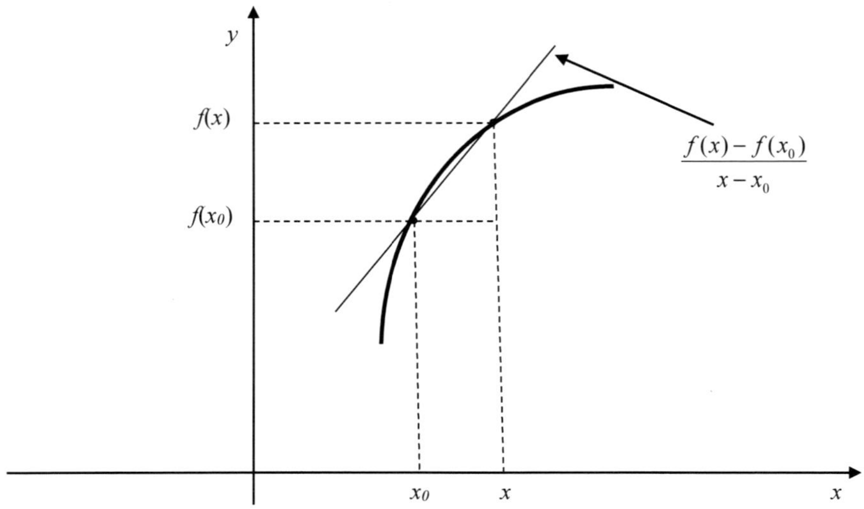

Figure 5.1. Geometric representation of the incremental ratio.

Let:

$$f : A \to \boldsymbol{R}, \qquad A \subseteq \boldsymbol{R}, \qquad x_0 \in A \cap D(A)$$

We define the *derivative of the function f at point x_0* the limit, if it exists, of the incremental ratio:

$$\left(\frac{df}{dx}\right)_{x=x_0} \equiv f'(x_0) \equiv \lim_{x \to x_0} \frac{f(x)-f(x_0)}{x-x_0} \tag{5.3}$$

The derivative, thus defined, is also called the first derivative.

Let:

$$f : A \to \boldsymbol{R}, \qquad A \subseteq \boldsymbol{R}, \qquad x_0 \in A \cap D(A);$$

$$\exists f'(x_0) \in \boldsymbol{R} \qquad \Rightarrow \qquad f \text{ is continuous in } x_0. \tag{5.4}$$

The existence of the derivative implies continuity only if the derivative is a finite number.

5.1.1. Example

Let:

$$f : \boldsymbol{R} \to \boldsymbol{R}, \qquad f(x) = x^2, \qquad x_0 \in \boldsymbol{R}$$

We have:

$$\begin{aligned} &\lim_{x \to x_0} \frac{f(x) - f(x_0)}{x - x_0} = \lim_{x \to x_0} \frac{x^2 - x_0{}^2}{x - x_0} = \\ &= \lim_{x \to x_0} \frac{(x + x_0)(x - x_0)}{x - x_0} = \lim_{x \to x_0} (x + x_0) = 2x_0 \end{aligned} \tag{5.5}$$

The limit of the incremental ratio of the function x^2 at point x_0 is $2x_0$.

5.1.2. Continuity and Derivability of a Function

As seen before, if it exists the derivative of a function at a point and it is a finite number then the function is continuous at that point (Eq. 5.4):

$$\exists f'(x_0) \in \boldsymbol{R} \qquad \Rightarrow \qquad f \text{ is continuous at } x_0.$$

The opposite implication is not always true: the continuity of a function does not always imply the existence of the derivative.

Let (Eq. 3.16):

$$f : \boldsymbol{R} \to \boldsymbol{R}, \qquad f(x) = |x|,$$

$$|x| = \begin{matrix} x & , x \geq 0 \\ -x & , x \leq 0 \end{matrix}$$

The absolute value function is continuous at point 0:

$$\lim_{x \to 0} |0| = 0$$

One may wonder if:

$$\exists f'(0) \quad ?$$

The incremental ratio at point 0 is:

$$\frac{f(x)-f(x_0)}{x-x_0}=\frac{f(x)-f(0)}{x-0}=\frac{|x|-|0|}{x-0}=\frac{|x|}{x}$$

For the existence of the derivative at point 0, it must be checked whether it exists:

$$\lim_{x\to 0}\frac{|x|}{x}=\lim_{x\to 0^+}\frac{|x|}{x}=\lim_{x\to 0^-}\frac{|x|}{x}$$

We calculate the right and left limits:

$$\left.\begin{array}{l}\lim\limits_{x\to 0^+}\dfrac{|x|}{x}=1\\[2ex]\lim\limits_{x\to 0^-}\dfrac{|x|}{x}=-1\end{array}\right\} \Rightarrow \quad \nexists\lim_{x\to 0}\frac{|x|}{x}$$

There is no first derivative at point 0 of the absolute value function, although it is a continuous function at point 0.

5.2. Considerations on the Incremental Ratio and Derivative of a Function

Let:

$$f: A\to \boldsymbol{R}, \qquad A\subseteq \boldsymbol{R}, \qquad x_0\in A\cap D(A)$$

As seen before, we define the derivative of the function f at point x_0 the limit, if it exists, of the incremental ratio (eq. 5.3):

$$\left(\frac{df}{dx}\right)_{x=x_0} \equiv f'(x_0) \equiv \lim_{x \to x_0} \frac{f(x) - f(x_0)}{x - x_0}$$

The derivative thus defined is also called the first derivative.
If we define:

$$h = x - x_0 \quad \Rightarrow \quad x = x_0 + h \tag{5.6}$$

The incremental ratio becomes:

$$\frac{f(x_0 + h) - f(x_0)}{h} \tag{5.7}$$

If, instead of the point x_0, we consider a generic point $x \in A \cap D(A)$, the incremental ratio can be written as:

$$\frac{f(x + h) - f(x)}{h} \tag{5.8}$$

and the definition of the derivative becomes:

$$\left(\frac{df}{dx}\right) \equiv f'(x) \equiv \lim_{h \to 0} \frac{f(x + h) - f(x)}{h} \tag{5.9}$$

In calculating the limit, the variable is represented by the increment h and the derivative refers to the function and not to a specific point x_0.
The derivative is no longer a number, as seen before, but it is in turn a function.

5.3. Properties of Derivatives and Summary Table of Derivation Rules

Let:

$$f, g : A \to \boldsymbol{R}, \qquad A \subseteq \boldsymbol{R}$$

We assume:

$$\exists f'(x), g'(x) \in \boldsymbol{R} \quad , \quad \forall x \in A \tag{5.10}$$

The following properties apply:

- derivative of a sum:

$$(f+g)'(x) = f'(x) + g'(x) \tag{5.11a}$$

- derivative of a product:

$$(f \cdot g)'(x) = f'(x)g(x) + f(x)g'(x) \tag{5.11b}$$

- derivative of the reciprocal of a function:

$$g \neq 0, \quad \left(\frac{1}{g}\right)'(x) = -\frac{g'(x)}{g^2(x)} \tag{5.11c}$$

- derivative of the ratio between two functions:

$$g \neq 0, \quad \left(\frac{f}{g}\right)'(x) = \frac{f'(x)}{g(x)} + f(x)\left(\frac{1}{g}\right)'(x) = \\ = \frac{f'(x)g(x) - f(x)g'(x)}{g^2(x)} \tag{5.11d}$$

Let:

$$f : A \to B,\ g : B \to \boldsymbol{R}, A, B \subseteq \boldsymbol{R}, \qquad y = f(x), z = g(y) = g(f(x))$$
$$\exists f'(x), g'(y) \in \boldsymbol{R}, \qquad \forall x \in A, y \in B$$

derivative of a composite function:

$$(g(f))'(x) = g'(y) \cdot f'(x) = g'(f(x)) \cdot f'(x) \tag{5.11e}$$

Let:

$$f : A \to B, g : B \to A, \qquad A, B \subseteq \boldsymbol{R} \qquad y = f(x), x = g(y)$$

$$\exists f'(x), g'(y) \in \boldsymbol{R}, \qquad \forall x \in A, y \in B, \qquad f' \neq 0$$

derivative of an inverse function:

$$g'(y) = \frac{1}{f'(x)} \tag{5.11f}$$

5.4. Derivative of the Exponential Function

We consider the exponential function:

$$f(x) = e^x, \qquad f: \boldsymbol{R} \to \boldsymbol{R}$$

The derivative is given by:

$$\frac{d}{dx}\left(e^x\right) = \lim_{h \to 0} \frac{e^{x+h} - e^x}{h} \tag{5.12}$$

Given the properties of the exponential function, it holds (Eq. 4.22):

$$e^{x+h} = e^x \cdot e^h$$

thus:

$$\begin{aligned} &\lim_{h \to 0} \frac{e^{x+h} - e^x}{h} = \lim_{h \to 0} \frac{e^x \cdot e^h - e^x}{h} = \lim_{h \to 0} \frac{e^x\left(e^h - 1\right)}{h} = \\ &= \lim_{h \to 0} e^x \frac{e^h - 1}{h} = e^x \cdot \lim_{h \to 0} \frac{e^h - 1}{h} \end{aligned} \tag{5.13}$$

e^x is not affected by the limit, which is calculated for the variable $h \to 0$, therefore it acts as a constant and it is extracted from the limit calculation operation.

As seen in exercise 9.5:

$$\lim_{h \to 0} \frac{e^h - 1}{h} = 1$$

therefore the following applies:

$$\lim_{h \to 0} \frac{e^{x+h} - e^x}{h} = e^x \cdot \lim_{h \to 0} \frac{e^h - 1}{h} = e^x \tag{5.14}$$

The derivative of an exponential function is:

$$\frac{d}{dx}\left(e^x\right) = e^x \tag{5.15}$$

5.5. Derivative of the Logarithm Function

We consider the logarithm function:

$$f(x) = \log x, \qquad f : \boldsymbol{R}^+ \to \boldsymbol{R}$$

The derivative is given by:

$$\frac{d}{dx}(\log x) = \lim_{h \to 0} \frac{\log(x+h) - \log x}{h} \tag{5.16}$$

From the properties of the logarithm function (Eq. 4.36):

$$\log(a \cdot b) = \log a + \log b$$

We have:

$$\begin{aligned} &\log(x+h) - \log x = \log(x+h) + \log x^{-1} = \\ &= \log\left[(x+h)x^{-1}\right] = \log\left(1 + \frac{h}{x}\right) \end{aligned} \tag{5.17}$$

We do a variable substitution:

$$z = \frac{h}{x} \qquad \Rightarrow \qquad h = z \cdot x$$

So when:

$$h \to 0 \qquad \Rightarrow \qquad z \to 0$$

The limit becomes:

$$\lim_{h\to 0}\frac{\log(x+h)-\log x}{h}=\lim_{h\to 0}\frac{\log\left(1+\dfrac{h}{x}\right)}{h}=\lim_{z\to 0}\frac{\log(1+z)}{z\cdot x}= \tag{5.18}$$
$$=\lim_{z\to 0}\frac{1}{x}\cdot\frac{\log(1+z)}{z}=\frac{1}{x}\cdot\lim_{z\to 0}\frac{\log(1+z)}{z}$$

As seen in exercise 9.4:

$$\lim_{z\to 0}\frac{\log(1+z)}{z}=1$$

therefore the following applies:

$$\lim_{h\to 0}\frac{\log(x+h)-\log x}{h}=\frac{1}{x}\cdot\lim_{z\to 0}\frac{\log(1+z)}{z}=\frac{1}{x} \tag{5.19}$$

The derivative of the logarithm function is:

$$\frac{d}{dx}(\log x)=\frac{1}{x} \tag{5.20}$$

Another way to calculate the derivative of the logarithm function is to consider the derivative of its inverse function.

Let:

$$f: A\to B, g: B\to A, \qquad A,B\subseteq \boldsymbol{R}, \qquad y=f(x), x=g(y)$$

$$\exists f'(x), g'(y)\in \boldsymbol{R}, \qquad \forall x\in A, y\in B, \qquad f'\neq 0, g'\neq 0$$

The functions *f* and *g* are one the inverse function of the other.

The derivative of *g* is given by (derivative of an inverse function, Eq. 5.11e):

$$g'(y)=\frac{1}{f'(x)}$$

The derivative of *f* is given by (derivative of an inverse function, Eq. 5.11e):

$$f'(x)=\frac{1}{g'(y)}$$

The inverse function of the logarithm is the exponential function and, in the case of the natural logarithm, it holds (the basis is constituted by the Napier's number, Eq. 4.41):

$$y=\log x \qquad \Leftrightarrow \qquad e^y=x$$

The derivative is given by:

$$\frac{d}{dx}(\log x)=\frac{1}{\frac{d}{dy}(e^y)}=\frac{1}{e^y}=\frac{1}{x};\left(\frac{d}{dy}(e^y)=e^y, e^y=x\right) \tag{5.21}$$

5.6. Derivative of the Power Function

Let:

$$f:\boldsymbol{R}^+\to\boldsymbol{R}, \qquad f(x)=x^a, \qquad a\in\boldsymbol{R}$$

We want to calculate the derivative:

$$\frac{d}{dx}\left(x^a\right)$$

Given a function:

$$g:\mathcal{C}(f)\to\boldsymbol{R}, \qquad z=g(f(x))$$

Its derivative is (derivative of composite function, Eq. 5.11e):

$$\frac{d}{dx}(g(f(x))) = g'(y)f'(x) = g'(f(x))f'(x)$$

In the case that we can identify a function for which the derivative:

$$\frac{d}{dx}(g(f(x)))$$

can be calculated independently of the explicit form for a composite function, it would be obtained:

$$f'(x) = \frac{\frac{d}{dx}(g(f(x)))}{g'(f(x))} \tag{5.22}$$

In the present case, the codomain of the function:

$$f: \boldsymbol{R}^+ \to \boldsymbol{R}, \qquad f(x) = x^a, \qquad a \in \boldsymbol{R}$$

is

$$\mathcal{C}(f) = \boldsymbol{R}^+$$

We can consider the function:

$$g: \boldsymbol{R}^+ \to \boldsymbol{R}, \qquad g(y) = g(f(x)) = \log x^a, \qquad a \in \boldsymbol{R} \tag{5.23}$$

Its derivative is:

$$\frac{d}{dx}\log(x^a) = \frac{1}{x^a} \cdot \frac{d}{dx}(x^a) \tag{5.24}$$

It can be observed that (Eq. 4.37):

$$\log x^a = a \cdot \log x$$

and:

$$\frac{d}{dx}\left(\log\left(x^a\right)\right)=\frac{d}{dx}\left(a\cdot\log x\right)=a\frac{d}{dx}\left(\log x\right)=\frac{a}{x} \tag{5.25}$$

thus:

$$\frac{a}{x}=\frac{1}{x^a}\cdot\frac{d}{dx}\left(x^a\right) \tag{5.26}$$

from which we obtain:

$$\frac{d}{dx}\left(x^a\right)=\frac{a}{x}\cdot x^a=a\cdot x^{-1}\cdot x^a=a\cdot x^{a-1} \tag{5.27}$$

The derivative of a power function is:

$$\frac{d}{dx}\left(x^a\right)=ax^{a-1} \tag{5.28}$$

5.7. Derivative of Trigonometric Functions

5.7.1. Derivative of the Sine and Cosine Functions

The derivative of the sine function is (without proof):

$$\frac{d}{dx}\left(\sin x\right)=\cos x \tag{5.29}$$

The derivative of the cosine function is (without proof):

$$\frac{d}{dx}\left(\cos x\right)=-\sin x \tag{5.30}$$

5.7.2. Derivative of the Tangent Function

Let (Eq. 4.59, 4.61a-b):

$$f(x) = \tan x = \frac{\sin x}{\cos x}$$
$$f : \boldsymbol{R} - \left\{ \left(k + \frac{1}{2} \right) \pi; k \in I \right\} \rightarrow \boldsymbol{R}$$

We want to calculate the derivative:

$$\frac{d}{dx}(f(x)) = \frac{d}{dx}(\tan x) = \frac{d}{dx}\left(\frac{\sin x}{\cos x}\right)$$

The derivative is (derivative of the ratio between two functions Eq. 5.11c):

$$\frac{d}{dx}\left(\frac{\sin x}{\cos x}\right) = \frac{\frac{d}{dx}(\sin x) \cdot \cos x - \sin x \cdot \frac{d}{dx}(\cos x)}{\cos^2 x} = \tag{5.31}$$
$$= \frac{\cos x \cdot \cos x - \sin x \cdot (-\sin x)}{\cos^2 x} = \frac{\cos^2 x + \sin^2 x}{\cos^2 x} = \frac{1}{\cos^2 x}$$

Thus:

$$\frac{d}{dx}(\tan x) = \frac{1}{\cos^2 x} \tag{5.32}$$

5.7.3. Derivative of the Cotangent Function

Let (Eq. 4.60, 4.67 – 4.71):

$$f(x) = \cot x = \frac{\cos x}{\sin x}$$
$$f : \boldsymbol{R} - \{k\pi; k \in I\} \rightarrow \boldsymbol{R}$$

We want to calculate the derivative:

$$\frac{d}{dx}(f(x)) = \frac{d}{dx}(\cot x) = \frac{d}{dx}\left(\frac{\cos x}{\sin x}\right)$$

The derivative is (derivative of the ratio between two functions Eq. 5.11c):

$$\frac{d}{dx}\left(\frac{\cos x}{\sin x}\right)=\frac{\frac{d}{dx}(\cos x)\cdot\sin x-\cos x\cdot\frac{d}{dx}(\sin x)}{\sin^2 x}= \\ =\frac{-\sin x\cdot\sin x-\cos x\cdot\cos x}{\sin^2 x}=\frac{-\sin^2 x-\cos^2 x}{\sin^2 x}=\frac{-1}{\sin^2 x} \tag{5.33}$$

Thus:

$$\frac{d}{dx}(\cot x)=-\frac{1}{\sin^2 x} \tag{5.34}$$

5.8. De L'Hôpital's Rule

Let:

$f,g:A\rightarrow \boldsymbol{R}$, $A\subseteq \boldsymbol{R}$, A interval of $\boldsymbol{R}$

$f,g\in C_A$ the functions f and g are continuous in every point of A

$\exists f'(x), g'(x)\in \boldsymbol{R}, \quad g'(x)\neq 0, \quad \forall x\in A,\ x_0\in D(A)$

De l'Hôpital's rule applies when the limit of the ratio between the two functions is an indeterminate form:

$$\lim_{x\to x_0}\frac{f(x)}{g(x)}=\frac{0}{0} \tag{5.35}$$

$$\lim_{x\to x_0}\frac{f(x)}{g(x)}=\frac{\pm\infty}{\pm\infty} \tag{5.36}$$

In both cases, the De l'Hôpital rule states that, if:

$$\exists\lim_{x\to x_0}\frac{f'(x)}{g'(x)} \quad\Rightarrow\quad \lim_{x\to x_0}\frac{f(x)}{g(x)}=\lim_{x\to x_0}\frac{f'(x)}{g'(x)} \tag{5.37}$$

Chapter 6

STUDY OF A FUNCTION: POINTS OF MAXIMUM AND MINIMUM, POINTS OF INFLECTION

6.1. DEFINITION OF SECOND DERIVATIVE

Let:

$$f : A \to \boldsymbol{R}, \qquad \exists f'(x) \in \boldsymbol{R}, \qquad \forall x \in A \tag{6.1}$$

We consider the function that associates to each x the derivative of f:

$$x \to f'(x), \qquad f' : A \to \boldsymbol{R} \tag{6.2}$$

and one point:

$$x_0 \in A \cap D(A)$$

If there exists the derivative of the function f' at point x_0 (i.e., the derivative of the first derivative of f at point x_0), it is called the second derivative (or derivative of order 2) at point x_0:

$$\frac{d}{dx}\left(f'(x_0)\right) \equiv (f')'(x_0) = f''(x_0) \equiv \left(\frac{d^2 f}{dx^2}\right)_{x=x_0} \tag{6.3}$$

To define the 2nd derivative at a point, the function must have the 1st derivative in a neighborhood of that point.

Similarly, if:

$$\exists f''(x) \in \boldsymbol{R}, \qquad \forall x \in A \tag{6.4}$$

We consider the function that associates to each x the 2nd derivative of f:

$$x \rightarrow f''(x), \qquad f'': A \rightarrow \boldsymbol{R} \tag{6.5}$$

and we define the third derivative at point x_0:

$$\frac{d}{dx}\left(f''(x_0)\right) \equiv (f'')'(x_0) = f'''(x_0) \equiv \left(\frac{d^3 f}{dx^3}\right)_{x=x_0} \tag{6.6}$$

In a similar way, we can define the fourth derivative, the fifth derivative, etc.

6.2. Monotone Increasing Functions and Monotone Decreasing Functions in an Interval

Let:

$$f: A \rightarrow \boldsymbol{R}, \qquad A \subseteq \boldsymbol{R},\ A \text{ interval of } \boldsymbol{R}$$

We say that:

i) f is a monotone increasing function if:

$$x < x' \quad \Rightarrow \quad f(x) \leq f(x')$$

ii) f is a strictly monotone increasing function if:

$$x < x' \quad \Rightarrow \quad f(x) < f(x')$$

iii) f is a monotone decreasing function if:

$$x < x' \quad \Rightarrow \quad f(x) \geq f(x')$$

iv) f is a strictly monotone decreasing function if:

$$x < x' \quad \Rightarrow \quad f(x) > f(x')$$

6.2.1. Example

As example we consider the exponential function:

$$f(x) = a^x, \qquad x \in \boldsymbol{R}, a \in \boldsymbol{R}^+, \qquad a > 1 \quad or \quad a < 1$$

The two cases are represented in Figure 6.1.
We can observe that:

$$a^x, a > 1 \qquad x < x' \Rightarrow a^x < a^{x'}$$

The exponential function with $a > 1$ is strictly increasing.

$$a^x, a < 1 \qquad x < x' \Rightarrow a^x > a^{x'}$$

The exponential function with $a < 1$ is strictly decreasing.

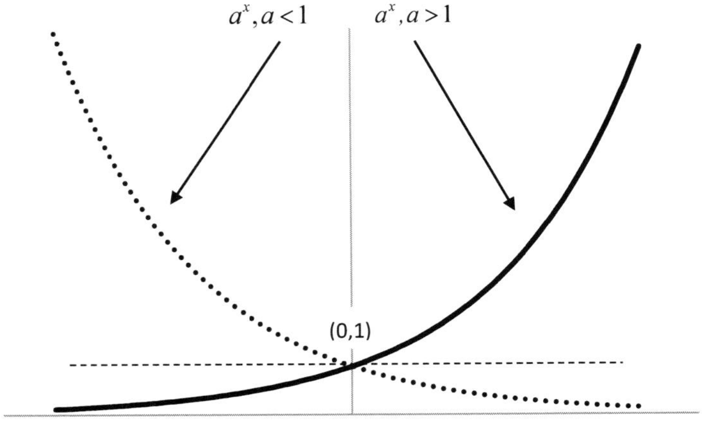

Figure 6.1. Graph of the exponential function: a^x, a >1 and a <1.

6.3. Increasing and Decreasing Functions at a Point

Let:

$f: A \rightarrow \boldsymbol{R}, \qquad A \subseteq \boldsymbol{R},\ A$ interval of $\boldsymbol{R},\ x_0 \in A$

We say that:
f is increasing at the point x_0 if:

$$\exists \delta > 0 : x \in A - \{x_0\}, |x - x_0| < \delta \quad \Rightarrow \quad \frac{f(x) - f(x_0)}{x - x_0} \geq 0 \quad (6.7a)$$

f is strictly increasing at the point x_0 if:

$$\exists \delta > 0 : x \in A - \{x_0\}, |x - x_0| < \delta \quad \Rightarrow \quad \frac{f(x) - f(x_0)}{x - x_0} > 0 \quad (6.7b)$$

The graph of an increasing function f must be with respect to a point x_0 in the areas graphically shown in Figure 6.2.

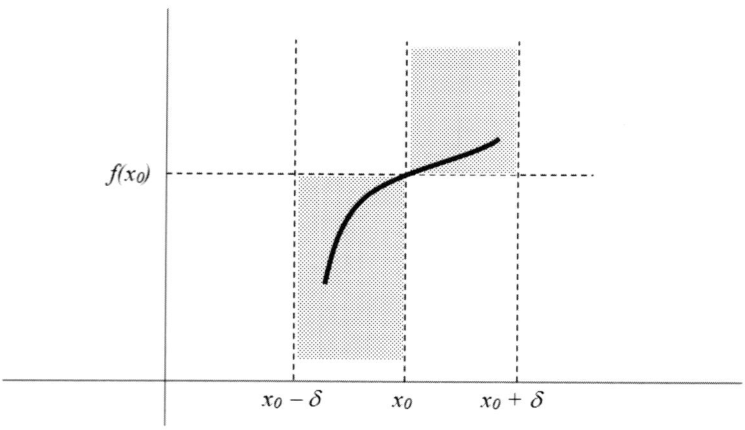

Figure 6.2. Graph of an increasing function at a point x_0.

The graph to the right and left of point x_0 will be found in the highlighted areas.
We say that:
f is decreasing at the point x_0 if:

$$\exists\delta > 0 : x \in A - \{x_0\}, |x - x_0| < \delta \quad \Rightarrow \quad \frac{f(x) - f(x_0)}{x - x_0} \leq 0 \tag{6.8a}$$

f is strictly increasing at the point x_0 if:

$$\exists\delta > 0 : x \in A - \{x_0\}, |x - x_0| < \delta \quad \Rightarrow \quad \frac{f(x) - f(x_0)}{x - x_0} < 0 \tag{6.8b}$$

The graph of a decreasing function *f* must be with respect to a point x_0 in the areas graphically shown in Figure 6.3.

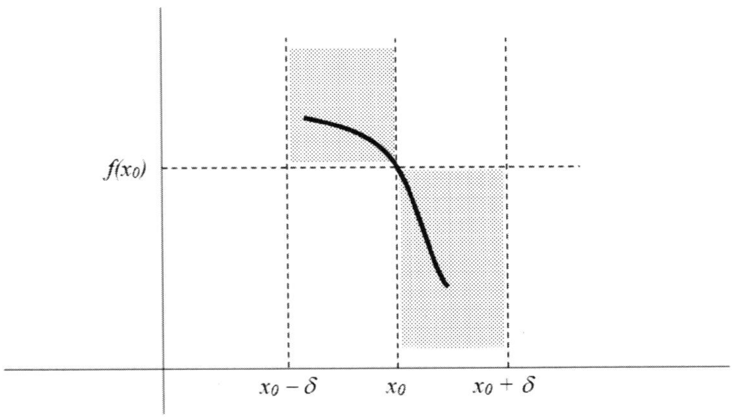

Figure 6.3. Graph of a decreasing function at a point x_0.

The graph to the right and left of point x_0 will be found in the highlighted areas.

It can be observed that the incremental ratio appears in the definition of increasing function and decreasing function.

So, if:

$$f : A \to \boldsymbol{R}, \qquad A \subseteq \boldsymbol{R},\ A \text{ interval of } \boldsymbol{R},\ x_0 \in A, \exists f'(x_0)$$

f is a strictly monotone increasing function if:

$$\frac{d}{dx} f(x_0) \equiv f'(x_0) > 0 \tag{6.9}$$

f is a strictly monotone decreasing function if:

$$\frac{d}{dx}f(x_0)\equiv f'(x_0)<0 \tag{6.10}$$

6.4. CONVEX AND CONCAVE FUNCTIONS

Let:

$$f:[a,b]\rightarrow \boldsymbol{R}$$

We consider the straight line segment with the extreme coordinates $x_1, f(x_1)$ and $x_2, f(x_2)$ represented in Figure 6.4:

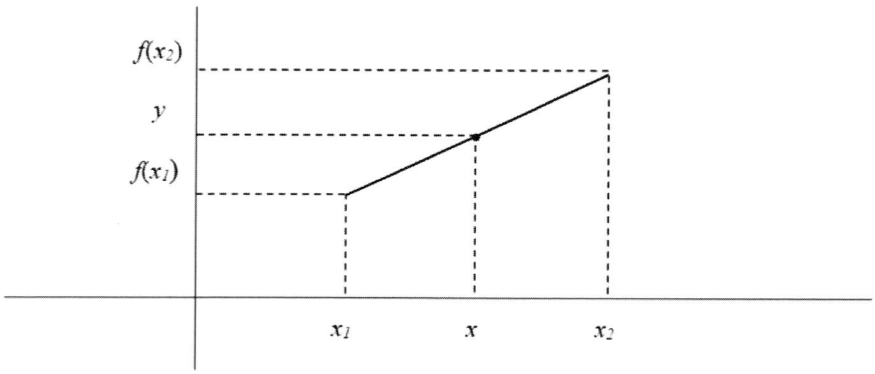

Figure 6.4. Graph of a straight line segment of coordinates $x_1, f(x_1)$ and $x_2, f(x_2)$.

A generic point x will be given by:

$$x = x_2 + t\cdot(x_1 - x_2) = t\cdot x_1 + (1-t)\cdot x_2, \quad t\in[0,1] \tag{6.11}$$

Consequently, a generic point y on the straight line segment will be given by:

$$y = t\cdot f(x_1) + (1-t)\cdot f(x_2) \tag{6.12}$$

A convex function is graphically represented in Figure 6.5.

The values of the function are lower than the y values of the straight line segment, so, for a generic point y, the following holds:

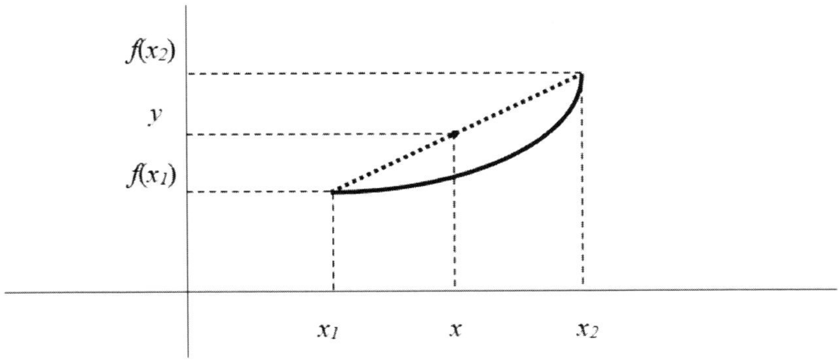

Figure 6.5. Graphical representation of a convex function.

$$y = f(x) = f(t \cdot x_1 + (1-t) \cdot x_2) \leq t \cdot f(x_1) + (1-t) \cdot f(x_2) \tag{6.13}$$

An example of a convex function is the exponential function.
A concave function is graphically represented in Figure 6.6.

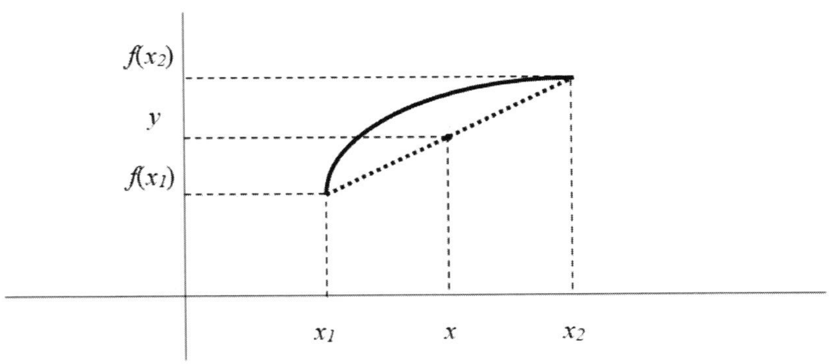

Figure 6.6. Graphical representation of a concave function.

The values of the function are greater than the y values of the straight line segment, so, for a generic point y, the following holds:

$$y = f(x) = f(t \cdot x_1 + (1-t) \cdot x_2) \geq t \cdot f(x_1) + (1-t) \cdot f(x_2) \tag{6.14}$$

An example of a concave function is the logarithm function.
Let:

$$f : [a,b] \to \boldsymbol{R} \quad , \quad \exists f', f''$$

We say that:

f is convex $\Leftrightarrow$

$$\forall x_1, x_2 \in [a,b] \quad , \quad \forall t \in [0,1]$$
$$f(t \cdot x_1 + (1-t) \cdot x_2) \leq t \cdot f(x_1) + (1-t) \cdot f(x_2) \tag{6.15}$$

f is concave $\Leftrightarrow$

$$\forall x_1, x_2 \in [a,b] \quad , \quad \forall t \in [0,1]$$
$$f(t \cdot x_1 + (1-t) \cdot x_2) \geq t \cdot f(x_1) + (1-t) \cdot f(x_2) \tag{6.16}$$

From the two definitions it can be shown that (the proof is not made):

f is convex $\Rightarrow f'(x)$ is an increasing function (6.17)

f is concave $\Rightarrow f'(x)$ is a decreasing function (6.18)

So, considering the derivative of the function $f'(x)$, we have:

$f''(x) > 0, \forall x \in [a,b] \Rightarrow f'(x)$ is increasing and $f(x)$ is convex (6.19)

$f''(x) < 0, \forall x \in [a,b] \Rightarrow f'(x)$ is decreasing and $f(x)$ is concave (6.20)

6.5. Maximum and Minimum of a Function

Let:

$A \subseteq \boldsymbol{R}$, A interval of $\boldsymbol{R}$

As seen previously (Eq. 2.14):

$A \subseteq \boldsymbol{R}$

has maximum if and only if:

$$\exists \sup A \qquad , \sup A \in A$$

The maximum is the limit superior when this belongs to the set (otherwise the set has no maximum). If the set has maximum:

$$\max A = \sup A$$

Similarly (Eq. 2.15):

$$A \subseteq \boldsymbol{R}$$

has minimum if and only if:

$$\exists \inf A \qquad , \inf A \in A$$

The minimum is the limit inferior when this belongs to the set (otherwise the set has no minimum). If the set has minimum:

$$\min A = \inf A$$

For example:

$$A =]0,1[\qquad 0 = \inf A, 0 \notin A \qquad 1 = \sup A, 1 \notin A$$

A has neither minimum nor maximum

$$A = [0,1] \qquad 0 = \inf A, 0 \in A \qquad 1 = \sup A, 1 \in A$$

A has minimum and maximum

To extend the notion of maximum and minimum to functions, we consider the codomain.

Let:

$$f : A \to \boldsymbol{R}, \qquad A \subseteq \boldsymbol{R}$$

Previously, the codomain of a function has been defined as (Eq. 3.4.):

$$f : A \to B$$
$$\mathcal{C}(f) = \{b \in B : \exists a \in A : b = f(a)\}$$

The codomain is the set of elements b belonging to B, with the property that there exists an element a belonging to A such that b is the image through f of a (expressed by $b = f(a)$).

The codomain, therefore, is the totality of the elements of B that have a point in A through the function f.

More briefly, the codomain can be defined as:

$$\mathcal{C}(f) = \{f(x); x \in A\} \tag{6.21}$$

If the codomain is superiorly bounded, it is said that the function f is upper bounded and this is indicated as:

$$\sup_A f = \sup \mathcal{C}(f) \tag{6.22}$$

If the codomain is inferiorly bounded, it is said that the function f is lower bounded and this is indicated as:

$$\inf_A f = \inf \mathcal{C}(f) \tag{6.23}$$

The function f has a maximum when the codomain has a maximum:

$$\exists \sup \mathcal{C}(f) \in \mathcal{C}(f) \quad \Rightarrow \quad \sup \mathcal{C}(f) = \max \mathcal{C}(f) \tag{6.24}$$

$$\max f = \sup_A f \equiv \sup \mathcal{C}(f) \tag{6.25}$$

The function f has a minimum when the codomain has a minimum:

$$\exists \inf \mathcal{C}(f) \in \mathcal{C}(f) \quad \Rightarrow \quad \inf \mathcal{C}(f) = \min \mathcal{C}(f) \tag{6.26}$$

$$\min f = \inf_A f \equiv \inf \mathcal{C}(f) \tag{6.27}$$

6.6. Points of Maximum and Minimum

Let:

$f : A \to \boldsymbol{R}, \qquad A \subseteq \boldsymbol{R}, \qquad x_0 \in A$

We define:

x_0 a point of *weak local maximum* of f if:

$$\exists \delta > 0 : x \in A - \{x_0\}, |x - x_0| < \delta \qquad \Rightarrow f(x) \le f(x_0) \tag{6.28}$$

x_0 a point of *strong local maximum* of f if:

$$\exists \delta > 0 : x \in A - \{x_0\}, |x - x_0| < \delta \qquad \Rightarrow f(x) < f(x_0) \tag{6.29}$$

x_0 a point of *weak local minimum* of f if:

$$\exists \delta > 0 : x \in A - \{x_0\}, |x - x_0| < \delta \qquad \Rightarrow f(x) \ge f(x_0) \tag{6.30}$$

x_0 a point of *strong local minimum* of f if:

$$\exists \delta > 0 : x \in A - \{x_0\}, |x - x_0| < \delta \qquad \Rightarrow f(x) > f(x_0) \tag{6.31}$$

In the graphic example of Figure 6.7, x_1 is a strong local maximum point and x_2 is a strong local minimum point, while the points of the interval $[x_3, x_4]$ are points of weak local maximum.

Let:

$f : A \to \boldsymbol{R}, \qquad A \subseteq \boldsymbol{R}$, A interval of $\boldsymbol{R}$,

$$x_0 \in \text{int}(A) \text{ (interior point of A)}, \ \exists f'(x_0) \tag{6.32}$$

If:

$$x_0 \text{ is a point of maximum} \ \Rightarrow \qquad f'(x_0) = 0 \tag{6.33}$$

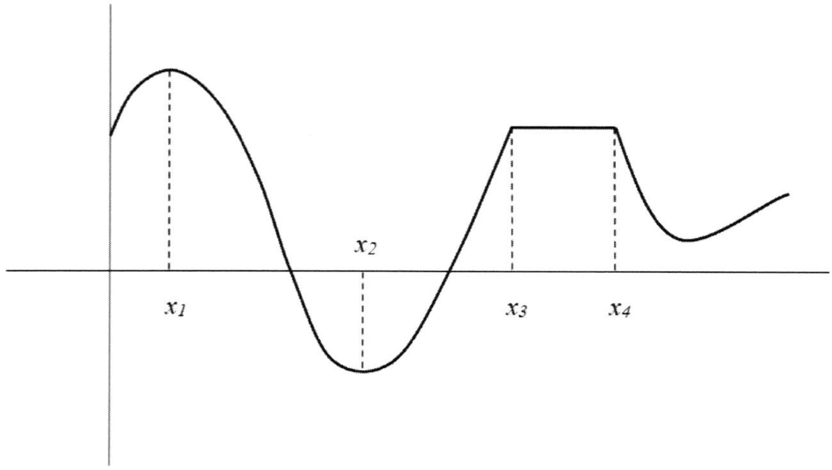

Figure 6.7. Graphical representation of local maximum and minimum points.

It can be easily verified with the following considerations.

If it were:

$$f'(x_0) > 0$$

The function would be increasing in x_0, so there would be points x to the right of x_0, in an arbitrary neighborhood of x_0, such that:

$$f(x) > f(x_0)$$

which contrasts with the fact that x_0 is the point of maximum.

If it were instead:

$$f'(x_0) < 0$$

The function would be decreasing in x_0, so there would be points x to the left of x_0, in an arbitrary neighborhood of x_0, such that:

$$f(x) > f(x_0)$$

which contrasts with the fact that x_0 is the point of maximum.

It can therefore be concluded that if:

$$x_0 \text{ is a } \textit{point of maximum}, \ \exists f'(x_0) \quad \Rightarrow \quad f'(x_0) = 0 \tag{6.34}$$

Similarly, for a point of minimum:

x_0 is a point of maximum $\Rightarrow \qquad f'(x_0)=0$ (6.35)

Even in this case, if it were:

$$f'(x_0)>0$$

The function would be increasing in x_0, so there would be points x to the left of x_0, in an arbitrary neighborhood of x_0, such that:

$$f(x)<f(x_0)$$

which contrasts with the fact that x_0 is the point of minimum.

If it were instead:

$$f'(x_0)<0$$

The function would be decreasing in x_0, so there would be points x to the right of x_0, in an arbitrary neighborhood of x_0, such that:

$$f(x)<f(x_0)$$

which contrasts with the fact that x_0 is the point of minimum.

It can therefore be concluded that if:

x_0 is a *point of minimum*, $\exists f'(x_0) \quad \Rightarrow \quad f'(x_0)=0$ (6.36)

The fact that the first derivative is null is a necessary condition for the existence of a maximum or minimum point, but it is not sufficient.

It can be observed that:

- at a *point of maximum* the function is *concave*;
- at a *point of minimum* the function is *convex*.

This allows us to say that, let be given:

$A \subseteq \boldsymbol{R}$, A interval of $\boldsymbol{R}$,

$x_0 \in \text{int}(A)$ (interior point of A), $\exists f'(x_0), f''(x_0)$ (6.37)

If:

x_0 is a *point of maximum* $\Rightarrow \quad f'(x_0)=0, \quad f''(x_0)<0$ (6.38)

x_0 is a *point of minimum* $\Rightarrow \quad f'(x_0)=0, \quad f''(x_0)>0$ (6.39)

6.7. POINTS OF INFLECTION

Let:

$$f:[a,b] \rightarrow \boldsymbol{R} \quad , \quad \exists f', f''$$

Specifically, we assume:

$$A=[a,b],\ x_0 \in \text{int}(A),\ \exists f'(x_0), f''(x_0) \tag{6.40}$$

We define x_0 point of inflection if:

- to the left of point x_0 the function is *concave*, to the right the function is *convex*;
- to the left of point x_0 the function is *convex*, to the right the function is *concave*.

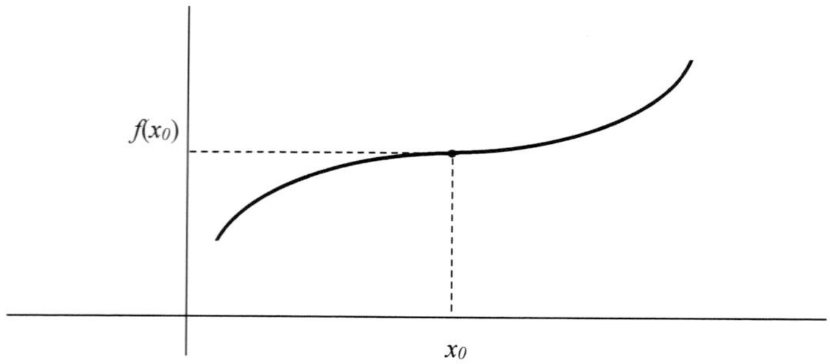

Figure 6.8a. Graphical representation of an inflection point: concave function on the left, convex on the right.

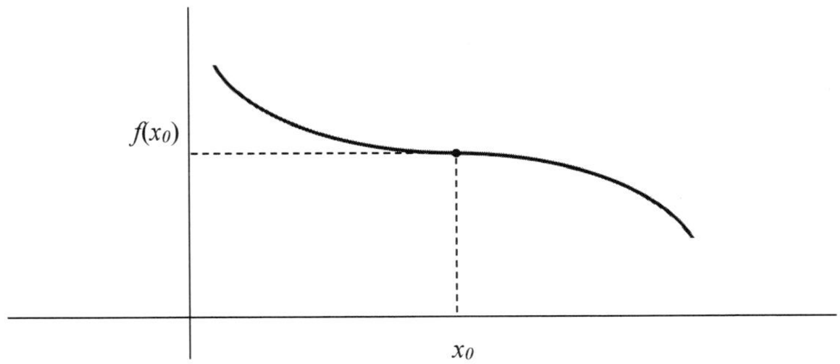

Figure 6.8b. Graphical representation of an inflection point: convex function on the left, concave on the right.

A graphical representation of the different curvatures of a function at an inflection point is given in Figures 6.8a and 6.8b.

This allows us to say that x_0 is an inflection point, if:

$f''(x_0)=0$, $f''(x)<0$ to the left of x_0, $f''(x)>0$ to the right of x_0;
$f''(x_0)=0$, $f''(x)>0$ to the left of x_0, $f''(x)<0$ to the right of x_0.

6.8. Asymptotes of a Function

Let:

$$f:A\to \boldsymbol{R} \quad , \quad A\subseteq \boldsymbol{R}$$

6.8.1. Vertical Asymptotes

The function has a *vertical asymptote* at the line $x = x_0$ parallel to the y axis, if:

$$\lim_{x\to x_0} f(x)=\pm\infty\,, x_0\in D(A) \tag{6.41}$$

Graphically, the cases that can be found are represented in Figure 6.9 (at most two simultaneously among the four represented).

In particular, the following cases may occur:

i) $\lim_{x\to x_0+} f(x)=+\infty\ (-\infty)$

$\lim_{x\to x_0-} f(x)=+\infty\ (-\infty)$

ii) $$\lim_{x \to x_0 +} f(x) = +\infty \ (-\infty)$$

$$\lim_{x \to x_0 -} f(x) = -\infty \ (+\infty)$$

The function graph shown in Figure 6.10 presents two infinite branches (continuous and dashed lines), both in the same half-plane with respect to the x axis, which have the same asymptote $x = x_0$.

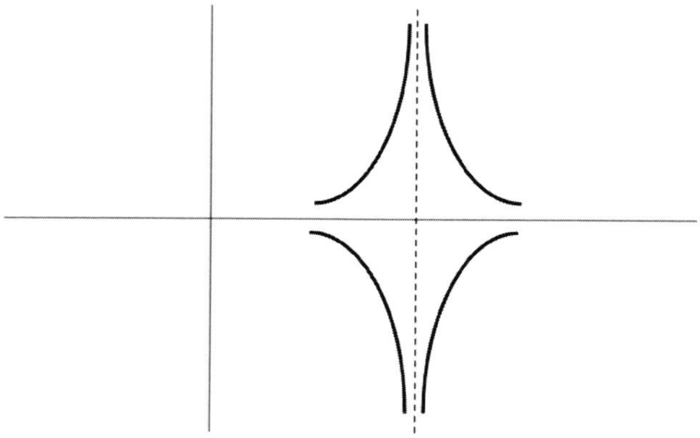

Figure 6.9. Graphical representation of all possible vertical asymptotes (not simultaneously).

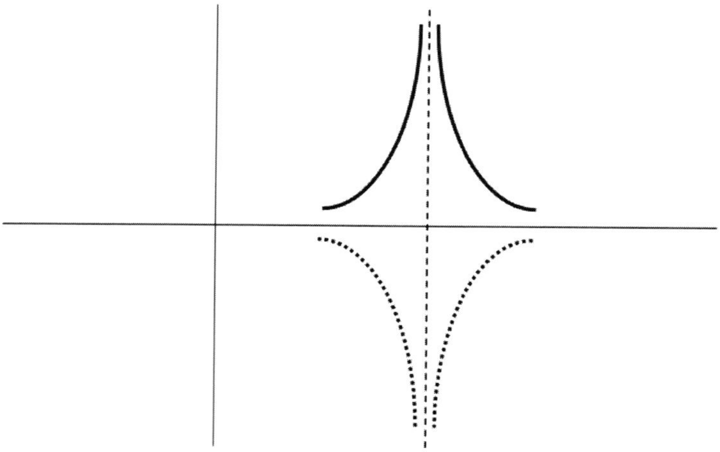

Figure 6.10. Graphical representation of vertical asymptotes with two infinite branches, both $+\infty$ or $-\infty$.

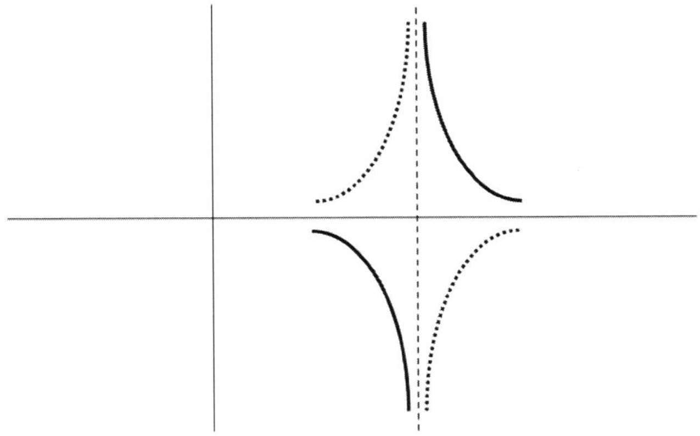

Figure 6.11. Graphical representation of vertical asymptotes with two infinite branches, $+\infty$ and $-\infty$ or $-\infty$ and $+\infty$.

The function graph shown in Figure 6.11 presents two infinite branches (continuous and dashed lines), one for each half-plane ($y > 0$, $y < 0$) with respect to the x axis, which have the same asymptote $x = x_0$.

iii) The function can have only one infinite branch that has asymptote $x = x_0$.

6.8.2. Horizontal Asymptotes

The function has a *horizontal asymptote* at the line $y = l$ parallel to the x axis, if:

$$\lim_{x \to \pm\infty} f(x) = l \, , l \in \boldsymbol{R} \tag{6.42}$$

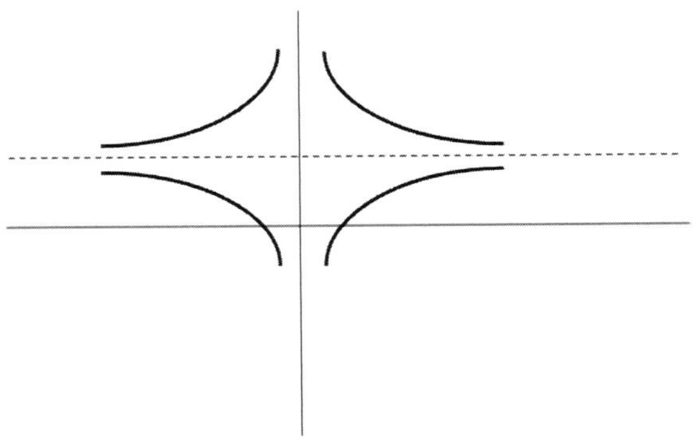

Figure 6.12. Graphical representation of all possible horizontal asymptotes (not simultaneously).

Graphically, the cases that can be found are represented in Figure 6.12 (at most two simultaneously among the four represented).

In particular, the following cases may occur:

i) The function has two infinite branches ($y \to +\infty$, $y \to -\infty$), both in the same half-plane with respect to the line $y = l$, which have the same asymptote $y = l$ (Figure 6.13, continuous and dashed lines).
ii) The function has two infinite branches ($y \to +\infty$, $y \to -\infty$), one for each half-plane with respect to the line $y = l$, which have the same asymptote $y = l$ (Figure 6.14, continuous and dashed lines).
iii) The function can have only one infinite branch which has asymptote $y = l$.

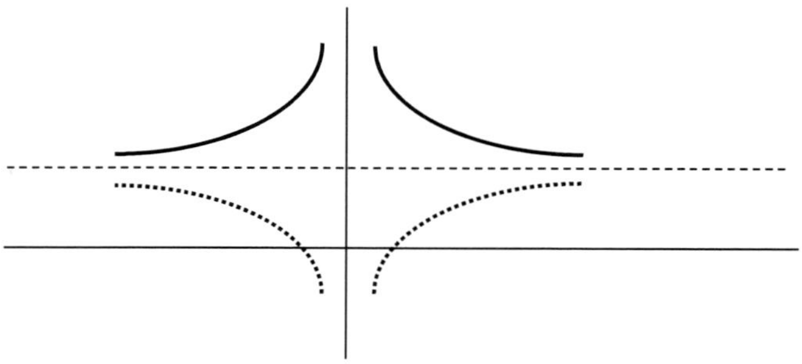

Figure 6.13. Graphical representation of horizontal asymptotes with two infinite branches, both in the same half-plane with respect to the line $y = l$.

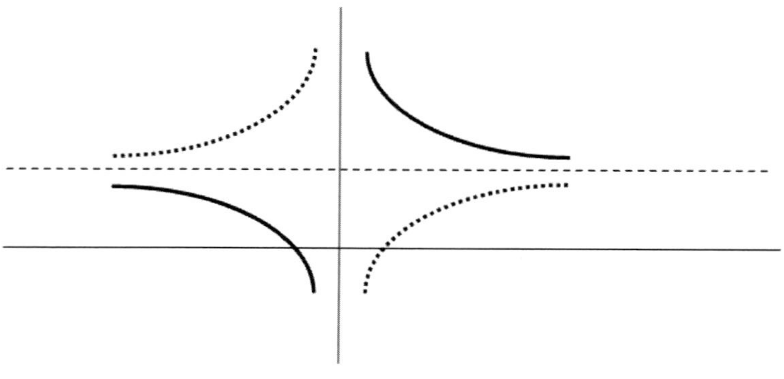

Figure 6.14. Graphical representation of horizontal asymptotes with two infinite branches, one for each half-plane with respect to the line $y = l$.

6.8.3. Oblique Asymptotes

If:

$$\lim_{x \to +\infty} f(x) = \pm\infty$$
$$\lim_{x \to -\infty} f(x) = \pm\infty \tag{6.43}$$

There are two possible cases:

i) The function does not have asymptotes.
ii) The function has an *oblique asymptote* at the straight line:

$$y = mx + q \tag{6.44}$$

where m is the *slope* of the straight line and q is the *y-intercept*.

Graphically, the cases that can be found are represented in Figures 6.14 and 6.15 (at most two simultaneously among the four cases represented for each figure).

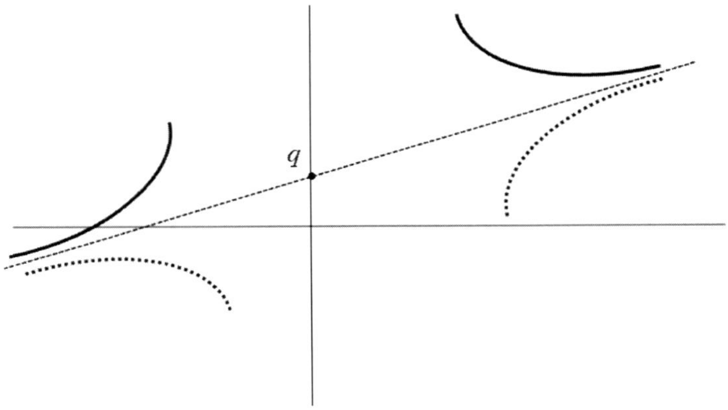

Figure 6.15. Graphical representation of oblique asymptotes with two infinite branches, both in the same half-plane with respect to the straight line of Eq. 6.44.

The function has an oblique asymptote at the straight line:

$$y = mx + q$$

if:

$$\lim_{x \to +\infty} \frac{f(x)}{x} = m, \ \lim_{x \to +\infty} [f(x) - mx] = q, \ m, q \in \boldsymbol{R} \tag{6.45}$$

$$\lim_{x \to -\infty} \frac{f(x)}{x} = m, \ \lim_{x \to -\infty} [f(x) - mx] = q, \ m, q \in \boldsymbol{R} \tag{6.46}$$

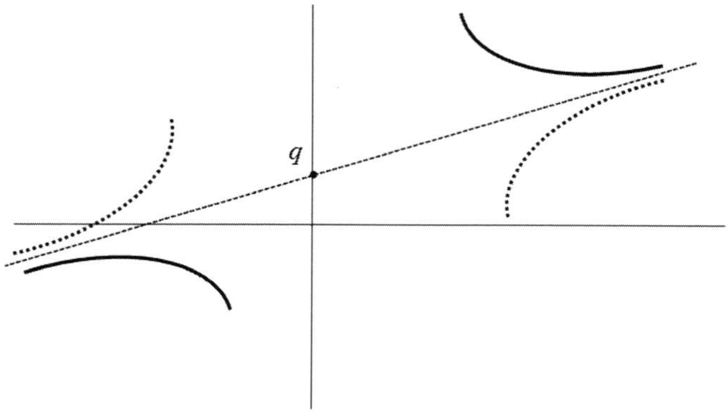

Figure 6.16 Graphical representation of oblique asymptotes with two infinite branches, one for each half-plane with respect to the straight line of Eq. 6.44.

In fact, if the function has an asymptote at the straight line, it must be:

$$\begin{aligned}&\lim_{x\to+\infty} f(x)=\lim_{x\to+\infty}(mx+q) \quad \Rightarrow\\&\lim_{x\to+\infty}\frac{f(x)}{x}=\lim_{x\to+\infty}\left(m+\frac{q}{x}\right)=m+\lim_{x\to+\infty}\frac{q}{x}=m\end{aligned} \tag{6.47}$$

$$\begin{aligned}&\lim_{x\to+\infty} f(x)=\lim_{x\to+\infty}(mx+q) \quad \Rightarrow\\&\lim_{x\to+\infty} f(x)-\lim_{x\to+\infty} mx=\lim_{x\to+\infty}[f(x)-mx]=\lim_{x\to+\infty} q=q\end{aligned} \tag{6.48}$$

A similar condition holds for $x \to -\infty$ (both or only one can occur).
If:

$$\lim_{x\to+\infty}\frac{f(x)}{x}=\pm\infty,\ \lim_{x\to-\infty}\frac{f(x)}{x}=\pm\infty \tag{6.49}$$

the function has no oblique asymptotes.

Even if:

$$\lim_{x\to+\infty}\frac{f(x)}{x}=0,\ \lim_{x\to-\infty}\frac{f(x)}{x}=0 \tag{6.50}$$

the function has no oblique asymptotes.

In fact $m = 0$ corresponds to a horizontal straight line, which indicates a horizontal asymptote that should have already been identified (Eq. 6.42). If we got here, it means that the horizontal asymptote is not there and there is not even an oblique asymptote.

Chapter 7

INDEFINITE INTEGRAL

7.1. INFINITESIMALS AND DIFFERENTIALS

7.1.1. Infinitesimals

A function that tends to 0 for $x \to x_0$ ($x \to \pm\infty$), is defined *infinitesimal* for $x \to x_0$ ($x \to \pm\infty$).

Let:

$$f, g : A \to \mathbf{R} \quad , \quad A \subseteq \mathbf{R}$$

and:

$$x_0 \in D(A)$$

or:

$x_0 \in D(A) \cap A$, x_0 accumulation point belonging to A

and/or:

$$+\infty \in D(A)$$
$$-\infty \in D(A)$$

We define:

$\lim_{x \to x_0} f(x) = 0$, *f* infinitesimal for $x \to x_0$

$\lim_{x \to x_0} g(x) = 0$, *g* infinitesimal for $x \to x_0$

$\lim_{x \to +\infty} f(x) = 0$, *f* infinitesimal for $x \to +\infty$

$\lim_{x \to -\infty} f(x) = 0$, *f* infinitesimal for $x \to -\infty$

$\lim_{x \to +\infty} g(x) = 0$, *g* infinitesimal for r $x \to +\infty$

$\lim_{x \to -\infty} g(x) = 0$, *g* infinitesimal for $x \to -\infty$

7.1.1.1. Order of an Infinitesimal

Given the two functions *f* and *g*, *f* is defined as an *infinitesimal of higher order* than *g* if:

$$\lim_{x \to x_0} \frac{f(x)}{g(x)} = 0$$

$$\lim_{x \to \pm\infty} \frac{f(x)}{g(x)} = 0$$

We can say that *f* approximates zero more rapidly than *g*.

Given the two functions *f* and *g*, *f* is defined as an *infinitesimal of lower order* than *g* if:

$$\lim_{x \to x_0} \frac{f(x)}{g(x)} = \pm\infty$$

$$\lim_{x \to \pm\infty} \frac{f(x)}{g(x)} = \pm\infty$$

We can say that *f* approximates zero less rapidly than *g*.

Given the two functions *f* and *g*, *f* and *g* are defined as *infinitesimals of the same order* if:

$$\lim_{x \to x_0} \frac{f(x)}{g(x)} = l \neq 0$$

$$\lim_{x \to \pm\infty} \frac{f(x)}{g(x)} = l \neq 0$$

We can say that *f* and *g* tend to zero with the same rapidity.

7.1.2. Differentials

Let:

$f:[a,b] \to \boldsymbol{R}$, *f* derivable in a generic point $x \in [a,b]$

Given the definition of derivative (Eq. 5.9):

$$\left(\frac{df}{dx}\right) \equiv f'(x) = \lim_{h \to 0} \frac{f(x+h) - f(x)}{h}$$

We have:

$$\frac{f(x+h) - f(x)}{h} - f'(x) \underset{h \to 0}{\to} 0 \tag{7.1}$$

The difference can be written:

$$\frac{f(x+h) - f(x)}{h} - f'(x) = \sigma(h) \tag{7.2}$$

Wit $\sigma(h)$ infinitesimal function: $\sigma(h) \underset{h \to 0}{\to} 0$.

We can write:

$$f(x+h) - f(x) = f'(x) \cdot h + \sigma(h) \cdot h \tag{7.3}$$

The increase Δy of the function:

$$\Delta y = f(x+h) - f(x) \tag{7.4}$$

it is decomposed into the sum:

$$\Delta y = f'(x) \cdot h + \sigma(h) \cdot h \tag{7.5}$$

Where:

$f'(x) \cdot h$ represents the *differential* of the function $f(x)$ in a generic point x.
$\sigma(h) \cdot h$ represents the *remainder* or *complementary part* of the function $f(x)$.

The differential is referred to as:

$$dy = df(x) = f'(x) \cdot h \tag{7.6}$$

A function for which it holds:

$$\Delta y = f'(x) \cdot h + \sigma(h) \cdot h \tag{7.7}$$

it is called *differentiable* at a generic point x.

If a function $f(x)$ is derivable in a generic point x it is also differentiable in the same point and vice versa.

If:

$$f(x) = x \tag{7.8}$$

we have:

$$dx = f'(x) \cdot h = 1 \cdot h = h \tag{7.9}$$

The differential of the independent variable x is equal to the increase h.

Therefore, the equation:

$$dy = f'(x) \cdot h$$

it can be written:

$$dy = f'(x) \cdot dx \tag{7.10}$$

from which:

$$\frac{dy}{dx} = f'(x) \tag{7.11}$$

The derivative $f'(x)$ is the ratio between the differential of the function and the differential of the independent variable.

7.2. Indefinite Integral

Let:

$$f:[a,b] \to \boldsymbol{R}, \qquad f \in C_{[a,b]} \tag{7.12}$$

We define *primitive* of $f(x)$ any function F:

$$F:[a,b] \to \boldsymbol{R}: \quad F'(x) = f(x) \qquad \forall x \in [a,b] \tag{7.13}$$

That is, a function that has $f(x)$ as a derivative on the whole definition interval.
If a function admits one primitive, it admits infinite primitives that differ by a constant:

$$F(x) = \Phi(x) + c \tag{7.14}$$

$$F'(x) = \Phi'(x) \tag{7.15}$$

$$F'(x) = f(x) \quad \Rightarrow \quad \Phi'(x) = f(x) \tag{7.16}$$

The set of all primitive of $f(x)$ is defined *indefinite integral* of $f(x)$ and is denoted as:

$$\int f(x)dx \tag{7.17}$$

Where $f(x)$ is called the *integrand function*, i.e., the function being integrated.
It applies:

$$F'(x) = \frac{dF}{dx} \tag{7.18}$$

$$dF(x) = F'(x)dx = f(x)dx \tag{7.19}$$

$$\int dF(x) = \int F'(x)dx = \int f(x)dx = F(x) + c \tag{7.20}$$

7.2.1. Examples of Calculation of Some Indefinite Integrals

7.2.1.1. Example 1

Calculate the integral:

$$\int e^x dx \tag{7.21}$$

A primitive of e^x is:

$$F(x) = e^x \quad \rightarrow \quad F'(x) = e^x$$

The integral is:

$$\int e^x dx = e^x + c \tag{7.22}$$

7.2.1.2. Example 2

Calculate the integral:

$$\int \cos x dx \tag{7.23}$$

A primitive of $\cos x$ is:

$$F(x) = \sin x \quad \rightarrow \quad F'(x) = \cos x$$

The integral is:

$$\int \cos x dx = \sin x + c \tag{7.24}$$

7.2.1.3. Example 3

Calculate the integral:

$$\int \sin x dx \tag{7.25}$$

A primitive of sinx is:

$$F(x) = -\cos x \quad \rightarrow \quad F'(x) = \sin x$$

The integral is:

$$\int \sin x dx = -\cos x + c \tag{7.26}$$

7.3. PROPERTIES OF INTEGRALS

Let:

$$f, g : [a,b] \rightarrow \boldsymbol{R} \qquad f, g \in C_{[a,b]}, \alpha \in \boldsymbol{R}, \alpha \neq 0 \tag{7.27}$$

α is a constant.
The following properties apply:

$$\int \alpha \cdot f(x) dx = \alpha \cdot \int f(x) dx \tag{7.28}$$

The integral of the product of a constant for a function is equal to the product of the constant for the integral of the function.

$$\int (f(x) + g(x)) dx = \int f(x) dx + \int g(x) dx \tag{7.29}$$

The integral of a sum of functions is equal to the sum of the integrals of the individual functions (distributive property of the integral with respect to the sum).

7.4. INTEGRATION BY SUBSTITUTION

Let:

$$f : [a,b] \rightarrow \boldsymbol{R} \quad , \quad f \in C_{[a,b]}$$

We want to calculate the integral:

$$\int f(x)dx$$

Let be identified:

$$\varphi:[t_1,t_2]\rightarrow \boldsymbol{R} \qquad \exists\varphi'(t), \qquad \varphi,\varphi'\in C_{[t_1,t_2]} \tag{7.30}$$

φ and φ' are continuous on the interval $[t_1,t_2]$
such that:

$$x=\varphi(t)\in[a,b] \quad ,\forall t\in[t_1,t_2] \tag{7.31}$$

from which it is noted:

$$\varphi:[t_1,t_2]\rightarrow[a,b] \tag{7.32}$$

The interval $[a,b]$ corresponds to the codomain of φ.
The differential dx is given by:

$$dx=\frac{dx}{dt}dt=\frac{d\varphi}{dt}dt=\varphi'(t)dt \tag{7.33}$$

The integral can thus be written as:

$$\int f(x)dx=\int f[\varphi(t)]\varphi'(t)dt \tag{7.34}$$

In the event that we do not know how to calculate one of the two integrals that appear in the Eq. 7.34, the substitution of a variable with the function of another variable can allow to calculate the integral.

7.4.1. Example of Calculation of an Integral by Substitution

Calculate the integral:

$$\int \frac{\log x}{x} dx \tag{7.35}$$

Considering the domain of definition of the logarithm function and the condition relating to the denominator, for the integrand function it holds:

$$f(x) = \frac{\log x}{x} \qquad f :]0,+\infty[\to \boldsymbol{R} \quad , f \in C_{]0,+\infty[} \tag{7.36}$$

It can be observed that:

$\dfrac{\log x}{x}$ does not have a primitive.

$$\frac{1}{x} = \frac{d}{dx}(\log x) \tag{7.37}$$

We can do a variable substitution:

$$t = \log x \tag{7.38}$$

from which:

$$dt = \frac{d}{dx}(\log x) \cdot dx = \frac{dx}{x} \tag{7.39}$$

The integral becomes:

$$\int \frac{\log x}{x} dx = \int t dt \tag{7.40}$$

The integral holds:

$$\int t dt = \frac{1}{2} t^2 + c \tag{7.41}$$

In general, for a power function it holds:

$$\frac{d}{dx}\left(x^a\right) = ax^{a-1} \quad \Rightarrow \quad x^{a-1} = \frac{1}{a}\cdot\frac{d}{dx}\left(x^a\right) = \frac{d}{dx}\left(\frac{x^a}{a}\right) \tag{7.42}$$

thus, for a power function the primitive is:

$$f(x) = F'(x) = x^a \quad \Rightarrow \quad F(x) = \frac{1}{a+1}\cdot x^{a+1} \tag{7.43}$$

So, considering the substitution made, the integral is (Eq. 7.60):

$$\int \frac{\log x}{x}dx = \frac{1}{2}\log^2 x + c \tag{7.44}$$

7.5. Integration by Parts

Let:

$$f, g : [a,b] \to \boldsymbol{R} \quad , \quad f, g \in C_{[a,b]} \tag{7.45}$$

We want to calculate:

$$\int f(x)\cdot g(x)dx \tag{7.46}$$

We assume that:

$$\exists f' \in C_{[a,b]} \tag{7.47}$$

f has continuous derivative over the entire definition interval.

$$\exists G : [a,b] \to \boldsymbol{R} \quad , \quad G'(x) = g(x) \qquad \forall x \in [a,b] \tag{7.48}$$

G is a primitive of g over the entire definition interval.
If we consider the product:

$$f(x)\cdot G(x) \tag{7.49}$$

Its differential is given by (Eq. 7.10):

$$d\left(f(x)\cdot G(x)\right)=\frac{d}{dx}\left(f(x)\cdot G(x)\right)dx \tag{7.50}$$

So the product differential of two functions results:

$$d\left(f(x)\cdot G(x)\right)=f(x)\cdot g(x)dx+f'(x)\cdot G(x)dx \tag{7.51}$$

By integrating we obtain:

$$\int d\left(f(x)\cdot G(x)\right)=\int f(x)\cdot g(x)dx+\int f'(x)\cdot G(x)dx \tag{7.52}$$

and the integral to be calculated (Eq.7.65) is:

$$\int f(x)\cdot g(x)dx=\int d\left(f(x)\cdot G(x)\right)-\int f'(x)\cdot G(x)dx \tag{7.53}$$

thus:

$$\int f(x)\cdot g(x)dx=f(x)\cdot G(x)-\int f'(x)\cdot G(x)dx \tag{7.54}$$

7.5.1. Examples of Calculation of Some Integrals by Parts

7.5.1.1. Example 1.

Calculate the integral:

$$\int xe^{x}dx \tag{7.55}$$

The greatest difficulty of the integration by parts consists in identifying which function allows the derivative and which the primitive.

In the present case it is assumed that:

$$f(x)=x \quad \rightarrow \quad f'(x)=1$$
$$g(x)=e^{x} \quad \rightarrow \quad G(x)=e^{x}$$

For integration by parts it applies (eq. 7.73):

$$\int f(x)\cdot g(x)dx = f(x)\cdot G(x) - \int f'(x)\cdot G(x)dx$$

Thus:

$$\int xe^x dx = xe^x - \int e^x dx = xe^x - e^x + c$$

The integral is therefore:

$$\int xe^x dx = e^x(x-1)+c \qquad (7.56)$$

7.5.1.2. Example 2

Calculate the integral:

$$\int \frac{\log x}{x} dx \qquad (7.57)$$

Previously (subsection 7.4.1), in defining the integration by substitution, this integral has been used as an example of calculation by substitution.

Specifically, the integral can be calculated both by substitution and by parts.

In the present case it is assumed that:

$$f(x) = \log x,\; g(x) = \frac{1}{x}$$

$$f'(x) = \frac{1}{x},\; G(x) = \log x$$

For integration by parts it applies:

$$\int f(x)\cdot g(x)dx = f(x)\cdot G(x) - \int f'(x)\cdot G(x)dx$$

for which:

$$\int \frac{\log x}{x} dx = \log x \cdot \log x - \int \frac{1}{x} \cdot \log x dx$$

If we collect the integrals:

$$\int \frac{\log x}{x} dx + \int \frac{\log x}{x} dx = \log x \cdot \log x$$

we obtain:

$$2\int \frac{\log x}{x} dx = \log^2 x$$

The integral is therefore:

$$\int \frac{\log x}{x} dx = \frac{1}{2} \log^2 x + c \tag{7.58}$$

That coincides with the result obtained with the integration by substitution.

7.5.1.3. Example 3

Calculate the integral:

$$\int \log x dx \tag{7.59}$$

It is assumed that:

$$f(x) = \log x, \ g(x) = 1$$
$$f'(x) = \frac{1}{x}, \ G(x) = x$$

By integrating by parts:

$$\int \log x dx = \log x \cdot x - \int \frac{1}{x} \cdot x dx = \log x \cdot x - \int dx =$$
$$= \log x \cdot x - x + c = x(\log x - 1) + c$$

The integral is therefore:

$$\int \log x dx = x(\log x - 1) + c \tag{7.60}$$

Chapter 8

DEFINITE INTEGRAL

8.1. DEFINITION OF THE DEFINITE INTEGRAL

Let:

$$f:[a,b]\rightarrow \boldsymbol{R} \quad , \quad f\in C_{[a,b]} \tag{8.1}$$

If the function is continuous in a finite interval, then the function is bounded. For example, the function is represented by the graph in Figure 8.1.

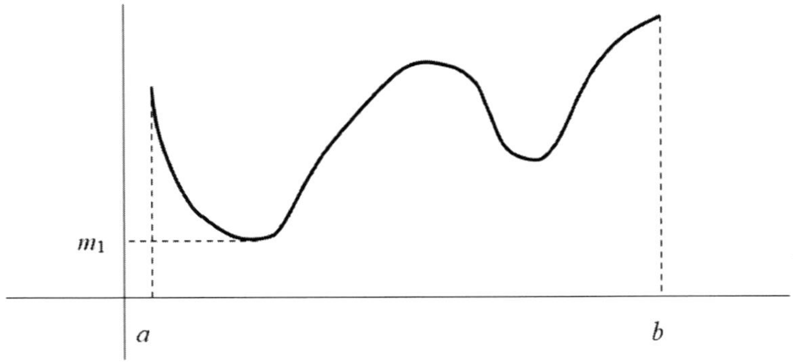

Figure 8.1. Graph of a generic function (Eq. 8.1).

The part of the plane delimited by the curve of the function, by the x axis and by the vertical segments $x = a$ and $x = b$ it is called *trapezoid*.

To calculate the area we can adopt a method that uses rectangles to approximate the trapezoid and such that the area of the rectangles, with a passage to the limit, is equal to the area of the trapezoid.

We can start by considering the rectangle based on the segment $[a,b]$ on the x axis and the minimum m_1 of the function as the height.

The area of the rectangle is given by:

$$S_1 = m_1(b-a) \tag{8.2}$$

We can then divide the interval $[a,b]$ into two equal parts δ_2 and consider the two rectangles which have as base δ_2 and as heights the two minimum m_1 and m_2 as shown in Figure 8.2.

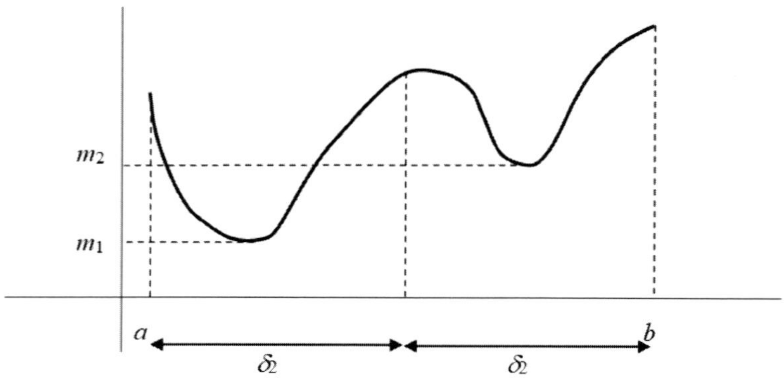

Figure 8.2. Graph of a generic function with two trapezoids indicated.

If we define:

$$\delta_2 = \frac{b-a}{2} \tag{8.3}$$

the area delimited by the two rectangles is:

$$S_2 = m_1\delta_2 + m_2\delta_2 \tag{8.4}$$

Carrying on, the interval can be divided into three equal parts by defining three rectangles with base δ_3 (Figure 8.3):

$$\delta_3 = \frac{b-a}{3} \tag{8.5}$$

and heights corresponding to the function minimums (if any) or to the lower values assumed by the function in the three intervals thus defined.

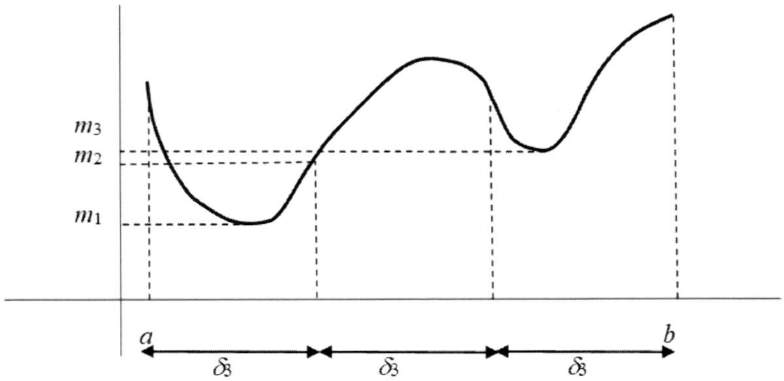

Figure 8.3. Graph of a generic function with three trapezoids indicated.

The area delimited by the three rectangles is:

$$S_3 = m_1\delta_3 + m_2\delta_3 + m_3\delta_3 \tag{8.6}$$

In general, by iterating the process and dividing the interval $[a,b]$ into n equal parts, we obtain:

$$\delta_n = \frac{b-a}{n} \tag{8.7}$$

$$S_n = \sum_{i=1}^{n} m_i \delta_n \tag{8.8}$$

The *definite integral* of the function f on the interval $[a,b]$ is defined as the limit:

$$\int_a^b f(x)dx = \lim_{n\to+\infty} \sum_{i=1}^{n} m_i \delta_n \tag{8.9}$$

where a and b represent the lower and upper limits of integration respectively.

The same result can be obtained by considering the maximum or average values of the function in the subdivision intervals, instead of the minimum values.

8.2. Properties of Definite Integrals

Let:

$$f, g : [a,b] \to \boldsymbol{R} \qquad f, g \in C_{[a,b]}, \qquad \alpha \in \boldsymbol{R}, \alpha \neq 0 \tag{8.10}$$

The same properties of indefinite integrals also apply to definite integrals:

$$\int_a^b \alpha \cdot f(x)dx = \alpha \cdot \int_a^b f(x)dx \tag{8.11}$$

$$\int_a^b \left(f(x) + g(x)\right)dx = \int_a^b f(x)dx + \int_a^b g(x)dx \tag{8.12}$$

Furthermore, let be given:

$$\begin{aligned} &f : [a,b] \to \boldsymbol{R} \quad , \quad f \in C_{[a,b]} \\ &x_1, x_2, x_3 \in [a,b] \quad , \quad x_1 \leq x_2 \leq x_3 \end{aligned} \tag{8.13}$$

It is worth (*additive property* of the definite integral):

$$\int_{x_1}^{x_3} f(x)dx = \int_{x_1}^{x_2} f(x)dx + \int_{x_2}^{x_3} f(x)dx \tag{8.14}$$

It also applies:

$$\int_a^b f(x)dx = -\int_b^a f(x)dx \tag{8.15}$$

8.2.1. Integral Mean

Let:

$$f : [a,b] \to \boldsymbol{R} \quad , \quad f \in C_{[a,b]}$$

We define *mean value* or *integral mean value* of the function *f* on the interval [*a*,b]:

$$\overline{f(x)} = \frac{1}{b-a}\int_a^b f(x)dx \tag{8.16}$$

The integral mean value is shown graphically in Figure 8.4.

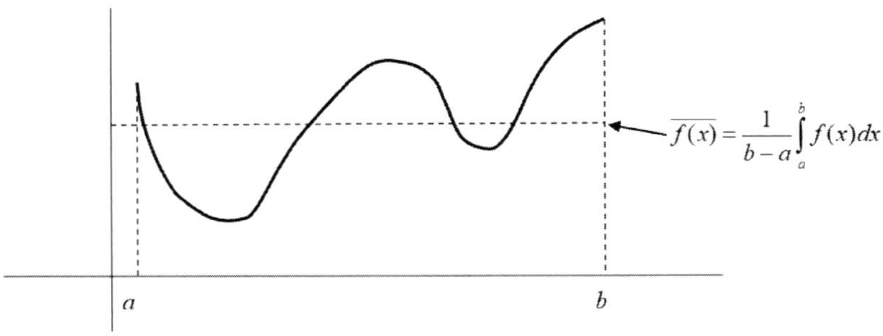

Figure 8.4. Graphical representation of the integral mean value.

The function, bounded on a finite interval, takes at least once a value corresponding to the mean value (four times in the example of Figure 8.4).

Considering a generic $\tau \in [a,b]$, which represents the value of the independent variable for which the function assumes a value corresponding to the mean value, we have:

$$\exists \tau \in [a,b]: \quad f(\tau) = \overline{f(x)} = \frac{1}{b-a}\int_a^b f(x)dx \quad \Rightarrow$$
$$\int_a^b f(x)dx = f(\tau)\cdot(b-a) \tag{8.17}$$

8.2.2. Integral Function

We can highlight an apparent difference between the indefinite integral and the definite integral:

- The indefinite integral involves the search for a function (the primitive) that has a derivative corresponding to the integrand function, so the indefinite integral *returns a function*.
- The definite integral involves the calculation of an area delimited by the curve of the integrand function and it *returns a value*.

Actually, the two notions of integral are linked together.

In fact, we can define the *integral function* by placing the variable x as the upper limit of integration:

$$F(x) = \int_a^x f(t)dt \tag{8.18}$$

If we want to calculate the derivative of the integral function $F(x)$ we can refer to the incremental ratio:

$$\frac{F(x+h) - F(x)}{h} = \frac{\int_a^{x+h} f(t)dt - \int_a^x f(t)dt}{h} \tag{8.19}$$

By the additive property of the definite integral (Eq. 8.14):

$$\frac{F(x+h) - F(x)}{h} = \frac{\int_a^{x+h} f(t)dt - \int_a^x f(t)dt}{h} = \\ = \frac{\int_a^x f(t)dt + \int_x^{x+h} f(t)dt - \int_a^x f(t)dt}{h} = \frac{\int_x^{x+h} f(t)dt}{h} \tag{8.20}$$

By the definition of integral mean (Eq. 8.17):

$$\frac{\int_x^{x+h} f(t)dt}{h} = \frac{[(x+h) - x] \cdot f(\tau)}{h} = \frac{h \cdot f(\tau)}{h} = f(\tau) \tag{8.21}$$

Thus:

$$\frac{F(x+h) - F(x)}{h} = \frac{\int_x^{x+h} f(t)dt}{h} = f(\tau) \tag{8.22}$$

For $h \to 0$, it can be observed that $\tau \to x$.

By performing the limit of the incremental ratio we obtain:

$$\lim_{h\to 0}\frac{F(x+h)-F(x)}{h}=F'(x)=\lim_{h\to 0}\frac{\int_x^{x+h} f(t)dt}{h}=f(x) \tag{8.23}$$

The integral function is a primitive of the integrand function:

$$F(x)=\int_a^x f(t)dt \quad\Rightarrow\quad F'(x)=f(x) \tag{8.24}$$

Given a primitive, there are infinite ones differing by a constant:

$$F(x)=\Phi(x)+c \tag{8.25}$$

It can be set:

$$c=-\Phi(a) \tag{8.26}$$

so it:

$$F(x)=\Phi(x)-\Phi(a)=\int_a^x f(t)dt \tag{8.27}$$

This allows to calculate a definite integral in a generic integration interval $[a,b]$:

$$\int_a^b f(t)dt=\Phi(b)-\Phi(a)=[\Phi(x)]_a^b=[F(x)]_a^b \tag{8.28}$$

8.3. Improper Integral

Previously (Eq. 8.1) we have considered the definite integral of a continuous function, therefore bounded, on a finite interval $[a,b]$.

For the definition of definite integral, the following two hypotheses have been made:

- Bounded function.
- Bounded interval of integration.

In many applications the two hypotheses are not verified, for example:

$$\int_{-\infty}^{+\infty} f(x)dx = 1$$

which may represent the case of an unbounded interval of integration for a probability density function.

Or:

$$\int_{0}^{1} \log x dx$$

where for the logarithm function:

$$f(x) = \log x \quad f : \boldsymbol{R}^{+} \rightarrow \boldsymbol{R}$$

$\lim_{x \to 0+} \log x = -\infty$ (unbounded function)

The following cases can be identified:

1. *Unbounded* functions and bounded integration intervals.
2. Bounded functions and *unbounded integration intervals.*
3. *Unbounded* functions and *unbounded integration intervals.*

It is therefore necessary to extend the notion of the definite integral to these cases.

8.3.1. Unbounded Function and Bounded Integration Interval

Let:

$$f : \left]a, b\right] \rightarrow \boldsymbol{R} \quad , \quad f \in C_{[\alpha, b]} \quad , \quad \alpha > a$$

$$\lim_{x \to a+} f(x) = \pm\infty \tag{8.29}$$

Under these hypotheses, the function is integrable on the interval $[\alpha, b]$:

$$\int_{\alpha}^{b} f(x)dx = l \quad , \quad l \in \boldsymbol{R} \tag{8.30}$$

It is said that if it exists the finite limit:

$$\lim_{\alpha \to a+} \int_{\alpha}^{b} f(x)dx \in \boldsymbol{R} \tag{8.31}$$

the function $f(x)$ is *integrable in a generalized sense* on the interval $[a, b]$ and the limit is called the *improper integral* or *generalized integral* of $f(x)$ on $[a, b]$:

$$\exists \lim_{\alpha \to a+} \int_{\alpha}^{b} f(x)dx \in \boldsymbol{R} \quad \Rightarrow \quad \int_{a}^{b} f(x)dx = \lim_{\alpha \to a+} \int_{\alpha}^{b} f(x)dx \tag{8.32}$$

Similar considerations apply to the upper limit of integration:

$$f : [a, b[\to \boldsymbol{R} \quad , \quad f \in C_{[a,\beta]} \quad , \quad \beta < b$$

$$\lim_{x \to b-} f(x) = \pm\infty \tag{8.33}$$

Under these hypotheses, the function is integrable on the interval $[a, \beta]$:

$$\int_{a}^{\beta} f(x)dx = l \quad , \quad l \in \boldsymbol{R} \tag{8.34}$$

$$\exists \lim_{\beta \to b-} \int_{a}^{\beta} f(x)dx \in \boldsymbol{R} \quad \Rightarrow \quad \int_{a}^{b} f(x)dx = \lim_{\beta \to b-} \int_{a}^{\beta} f(x)dx \tag{8.35}$$

8.3.2. Bounded Function and Unbounded Integration Interval

Let:

$$f:[a,+\infty[\to \boldsymbol{R} \quad , \quad f\in C_{[a,\beta]} \quad , \quad \beta>a \tag{8.36}$$

Under these hypotheses, the function is integrable on the interval $[a,\beta]$:

$$\int_a^\beta f(x)dx=l \quad , \quad l\in \boldsymbol{R} \tag{8.37}$$

It is said that if it exists the finite limit:

$$\lim_{\beta\to+\infty}\int_a^\beta f(x)dx\in \boldsymbol{R} \tag{8.38}$$

the function $f(x)$ is *integrable in a generalized sense* on the interval $[a,+\infty[$ and the limit is called the *improper integral* or *generalized integral* of $f(x)$ on $[a,+\infty[$:

$$\exists \lim_{\beta\to+\infty}\int_a^\beta f(x)dx\in \boldsymbol{R} \quad \Rightarrow \quad \int_a^{+\infty} f(x)dx=\lim_{\beta\to+\infty}\int_a^\beta f(x)dx \tag{8.39}$$

Similar considerations apply to the lower limit of integration:

$$f:]-\infty,b]\to \boldsymbol{R} \quad , \quad f\in C_{[\alpha,b]} \quad , \quad \alpha<b \tag{8.40}$$

Under these hypotheses, the function is integrable on the interval $[\alpha,b]$:

$$\int_\alpha^b f(x)dx=l \quad , \quad l\in \boldsymbol{R} \tag{8.41}$$

$$\exists \lim_{\alpha\to-\infty}\int_\alpha^b f(x)dx\in \boldsymbol{R} \quad \Rightarrow \quad \int_{-\infty}^{b} f(x)dx=\lim_{\alpha\to-\infty}\int_\alpha^b f(x)dx \tag{8.42}$$

8.3.3. Unbounded Function and Unbounded Integration Interval

Let:

$$f:]a,+\infty[\to \boldsymbol{R} \quad , \quad f \in C_{[\alpha,\beta]} \quad , \quad \alpha,\beta > a$$

$$\lim_{x\to a+} f(x) = \pm\infty \tag{8.43}$$

In this case, the definition of improper integral is obtained from the definitions given in the two previous cases (subsections 8.3.1 and 8.3.2).

Under these hypotheses, the function is integrable on the interval $[\alpha,\beta]$:

$$\int_{\alpha}^{\beta} f(x)dx = l \quad , \quad l \in \boldsymbol{R} \tag{8.44}$$

It is said that if finite limits exist:

$$\lim_{\alpha\to a+} \int_{\alpha}^{\beta} f(x)dx \in \boldsymbol{R} \ , \ \lim_{\beta\to+\infty} \int_{\alpha}^{\beta} f(x)dx \in \boldsymbol{R} \tag{8.45}$$

the function $f(x)$ is *integrable in a generalized sense* on the interval $]a,+\infty[$ and the double limit is called the *improper integral* or *generalized integral* of $f(x)$ on $]a,+\infty[$:

$$\begin{matrix} \exists \lim\limits_{\alpha\to a+} \int_{\alpha}^{\beta} f(x)dx \in \boldsymbol{R} \\ \exists \lim\limits_{\beta\to+\infty} \int_{\alpha}^{\beta} f(x)dx \in \boldsymbol{R} \end{matrix} \quad \Rightarrow \quad \int_{a}^{+\infty} f(x)dx = \lim_{\substack{\alpha\to a+ \\ \beta\to+\infty}} \int_{\alpha}^{\beta} f(x)dx \tag{8.46}$$

Similar considerations apply to the lower limit of integration:

$$f:]-\infty,b[\to \boldsymbol{R} \quad , \quad f \in C_{[\alpha,\beta]} \quad , \quad \alpha,\beta < b$$

$$\lim_{x\to b-} f(x) = \pm\infty \tag{8.47}$$

Under these hypotheses, the function is integrable on the interval $[\alpha,\beta]$:

$$\int_{\alpha}^{\beta} f(x)dx = l \quad , \quad l \in \boldsymbol{R} \tag{8.48}$$

$$\left.\begin{array}{l} \exists \lim\limits_{\beta \to b-} \int_{\alpha}^{\beta} f(x)dx \in \boldsymbol{R} \\ \exists \lim\limits_{\alpha \to -\infty} \int_{\alpha}^{\beta} f(x)dx \in \boldsymbol{R} \end{array}\right. \Rightarrow \quad \int_{-\infty}^{b} f(x)dx = \lim_{\substack{\beta \to b- \\ \alpha \to -\infty}} \int_{\alpha}^{\beta} f(x)dx \tag{8.49}$$

Chapter 9

CALCULATION OF FUNCTION LIMITS

9.1. PROBLEMS

For the following functions, find the domain and the specified limits.

9.1.1. Problem 9.1

$$f(x) = \frac{x}{1+x} \tag{9.1}$$

$$\lim_{x\to+\infty} \frac{x}{1+x}; \ \lim_{x\to-\infty} \frac{x}{1+x}; \ \lim_{x\to-1^-} \frac{x}{1+x}; \ \lim_{x\to-1^+} \frac{x}{1+x}$$

9.1.2. Problem 9.2

$$f(x) = \frac{a^x}{a^x + 1}, \qquad x \in \boldsymbol{R}, a \in \boldsymbol{R}^+ \tag{9.2}$$

$$\lim_{x\to 0} \frac{a^x}{a^x + 1}; \ \lim_{x\to+\infty} \frac{a^x}{a^x + 1}; \ \lim_{x\to-\infty} \frac{a^x}{a^x + 1}$$

9.1.3. Problem 9.3

$$f(x)=\frac{\log x}{1+\log x}\equiv\frac{\ln x}{1+\ln x} \tag{9.3}$$

$$\lim_{x\to 0+}\frac{\log x}{\log x+1};\ \lim_{x\to +\infty}\frac{\log x}{\log x+1};\ \lim_{x\to\left(e^{-1}\right)^{+}}\frac{\log x}{\log x+1};\ \lim_{x\to\left(e^{-1}\right)^{-}}\frac{\log x}{\log x+1}$$

9.1.4. Problem 9.4

$$f(x)=\frac{\log(x+1)}{x} \tag{9.4}$$

$$\lim_{x\to 0}\frac{\log(x+1)}{x}$$

9.1.5. Problem 9.5

$$f(x)=\frac{e^{x}-1}{x} \tag{9.5}$$

$$\lim_{x\to 0}\frac{e^{x}-1}{x}$$

9.1.6. Problem 9.6

$$f(x)=\frac{\sin x}{\sin x+1} \tag{9.6}$$

$$\lim_{x\to k\pi}\frac{\sin x}{\sin x+1};\ \lim_{x\to\left(\frac{1}{2}+2k\right)\pi}\frac{\sin x}{\sin x+1};\ \lim_{x\to\left(\frac{3}{2}+2k\right)\pi}\frac{\sin x}{\sin x+1}$$

9.1.7. Problem 9.7

$$f(x)=\frac{\sin x}{x} \tag{9.7}$$

$$\lim_{x\to 0}\frac{\sin x}{x};\ \lim_{x\to +\infty}\frac{\sin x}{x};\ \lim_{x\to -\infty}\frac{\sin x}{x}$$

9.1.8. Problem 9.8

$$f(x)=\frac{\log(3x+1)}{x} \tag{9.8}$$

$$\lim_{x\to 0}\frac{\log(3x+1)}{x}$$

9.1.9. Problem 9.9

$$f(x)=e^{-3x^2+2} \tag{9.9}$$

$$\lim_{x\to \pm\infty} e^{-3x^2+2}$$

9.1.10. Problem 9.10

$$f(x)=\frac{\log x}{x-1} \tag{9.10}$$

$$\lim_{x\to 1}\frac{\log x}{x-1}$$

9.1.11. Problem 9.11

$$f(x)=e^{4x^3-2x+1} \tag{9.11}$$

$$\lim_{x\to +\infty} e^{4x^3-2x+1}$$

9.1.12. Problem 9.12

$$f(x) = e^{4x^3 - 2x + 1} \tag{9.12}$$

$$\lim_{x \to -\infty} e^{4x^3 - 2x + 1}$$

9.1.13. Problem 9.13

$$f(x) = \frac{\sin 2x}{x} \tag{9.13}$$

$$\lim_{x \to 0} \frac{\sin 2x}{x}$$

9.1.14. Problem 9.14

$$f(x) = \frac{x - \sin x}{x + \sin x} \tag{9.14}$$

$$\lim_{x \to 0} \frac{x - \sin x}{x + \sin x}$$

9.1.15. Problem 9.15

$$f(x) = \frac{x - \sin x}{x + \sin x} \tag{9.15}$$

$$\lim_{x \to +\infty} \frac{x - \sin x}{x + \sin x}$$

9.1.16. Problem 9.16

$$f(x) = \frac{\left(1 - \sin^2 x\right)^2}{\cos x} \tag{9.16}$$

$$\lim_{x \to \frac{\pi}{2}} \frac{\left(1 - \sin^2 x\right)^2}{\cos x}$$

9.1.17. Problem 9.17

$$f(x) = \frac{1 - \cos x}{x^2} \tag{9.17}$$

$$\lim_{x \to 0} \frac{1 - \cos x}{x^2}$$

9.2. Solutions

9.2.1. Problem 9.1

Given the function:

$$f(x) = \frac{x}{1 + x}$$

Indicate the domain of definition of the function and calculate the limit:

$$\lim_{x \to +\infty} f(x) = \lim_{x \to +\infty} \frac{x}{1 + x}$$

The domain of definition of the function consists of the function domain in the numerator, in this case ***R***, minus the values in which the denominator is equal to zero: $1 + x = 0$.

It must therefore be verified:

$$1 + x \neq 0 \quad \Rightarrow \quad x \neq -1$$

The inequality found is not the domain, but it represents the condition that determines the domain.

The function is therefore defined by:

$$f : \boldsymbol{R} - \{-1\} \to \boldsymbol{R}$$

The function is defined on $\boldsymbol{R} - \{-1\}$, for which $+\infty$ is an accumulation point.

We can work on the expression of the function, trying to express it in terms of simpler functions whose limits are already known.

We consider:

$$\frac{x}{1+x} = \frac{1}{\frac{1}{x}\cdot(1+x)} = \frac{1}{\frac{1}{x}+1} \xrightarrow[x\to+\infty]{} = \frac{1}{0+1} = 1 \qquad \left(\frac{1}{x} \xrightarrow[x\to+\infty]{} 0\right)$$

The method followed to simplify the function is not unique, an alternative way is:

$$\frac{x}{1+x} = \frac{x}{x\cdot\left(\frac{1}{x}+1\right)} = \frac{1}{\frac{1}{x}+1} \xrightarrow[x\to+\infty]{} = \frac{1}{0+1} = 1 \qquad \left(\frac{1}{x} \xrightarrow[x\to+\infty]{} 0\right)$$

therefore:

$$\lim_{x\to+\infty} \frac{x}{1+x} = 1$$

In the same way:

$$\lim_{x\to-\infty} \frac{x}{1+x} = \lim_{x\to-\infty} \frac{1}{\frac{1}{x}+1} = \frac{1}{0+1} = 1$$

Given the domain of the function, we can calculate the limits:

$$\lim_{x\to-1^-} \frac{x}{1+x}$$

$$\lim_{x \to -1^+} \frac{x}{1+x}$$

We obtain:

$$\lim_{x \to -1^-} \frac{x}{1+x} = \frac{-1}{1-1^-} = \frac{-1}{0^-} = +\infty$$

$x \to -1^-$, then for values $x < -1$, so the denominator will tend to 0 from negative values, from which the ratio is $+\infty$.

$$\lim_{x \to -1^+} \frac{x}{1+x} = \frac{-1}{1-1^+} = \frac{-1}{0^+} = -\infty$$

$x \to -1^+$, then for values $x > -1$, so the denominator will tend to 0 from positive values, from which the ratio is $-\infty$.

The graph of the function is shown in Figure 9.1.

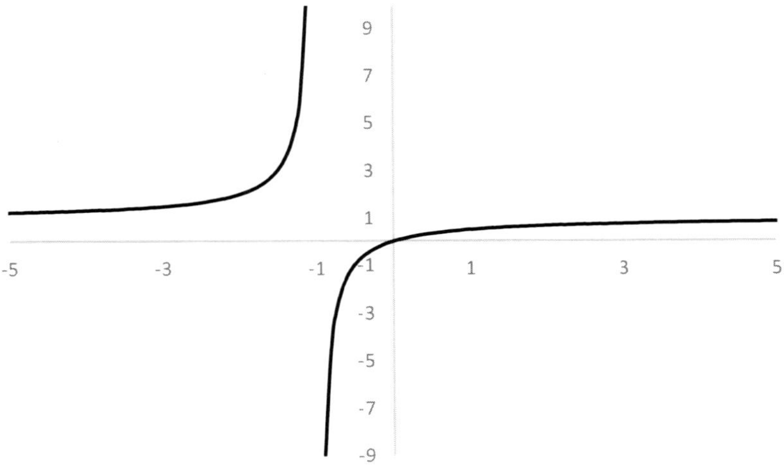

Figure 9.1. Graphical representation of the function of Eq. 9.1.

9.2.2. Problem 9.2

Let:

$$f(x) = \frac{a^x}{a^x + 1}, \qquad x \in \boldsymbol{R}, a \in \boldsymbol{R}^+$$

We want to calculate:

$$\lim_{x\to 0}\frac{a^x}{a^x+1};\ \lim_{x\to +\infty}\frac{a^x}{a^x+1};\ \lim_{x\to -\infty}\frac{a^x}{a^x+1}$$

The domain of definition of the function consists of the function domain in the numerator, in this case ***R***, minus the values in which the denominator is equal to zero.

Given the nature of the denominator:

$$a^x+1\neq 0$$

the domain of definition of the function is ***R***,

We have to distinguish the two cases:

a > 1

a < 1

In both cases:

$$\lim_{x\to 0}\frac{a^x}{a^x+1}=\frac{a^0}{a^0+1}=\frac{1}{1+1}=\frac{1}{2}$$

1st case: a > 1.

$$\lim_{x\to +\infty}\frac{a^x}{a^x+1}=\lim_{x\to +\infty}\frac{a^x}{a^x\left(1+\frac{1}{a^x}\right)}=\lim_{x\to +\infty}\frac{1}{1+\frac{1}{a^x}}=\lim_{x\to +\infty}\frac{1}{1+a^{-x}}=\frac{1}{1+0}=1$$

$$\lim_{x\to -\infty}\frac{a^x}{a^x+1}=\frac{0}{0+1}=\frac{0}{1}=0$$

2nd case: a < 1.

$$\lim_{x\to +\infty}\frac{a^x}{a^x+1}=\frac{0}{0+1}=\frac{0}{1}=0$$

$$\lim_{x\to -\infty}\frac{a^x}{a^x+1}=\lim_{x\to -\infty}\frac{1}{1+\frac{1}{a^x}}=\frac{1}{1+0}=1$$

9.2.3. Problem 9.3

Let:

$$f(x) = \frac{\log x}{1+\log x} \equiv \frac{\ln x}{1+\ln x}$$

The function definition domain is given by the intersection of the function domains in the numerator and the denominator, minus the values in which the denominator is equal to 0.

Both the numerator and the denominator have the logarithm function, defined by:

$$\log : \boldsymbol{R}^{+} \to \boldsymbol{R}$$

From the function definition domain it must be removed the points in which the denominator is equal to 0:

$$1+\log x = 0 \quad \Rightarrow \quad \log x = -1$$

The inverse function of the logarithm is used to solve the equation (Eq. 4.41):

$$e^{\log x} = e^{-1} \quad \Rightarrow \quad x = e^{-1}$$

Point e^{-1} is removed from the domain, so the function is defined by:

$$f : \boldsymbol{R}^{+} - \{e^{-1}\} \to \boldsymbol{R}$$

We want to calculate:

$$\lim_{x\to 0+} \frac{\log x}{\log x + 1}, \quad \lim_{x\to +\infty} \frac{\log x}{\log x + 1}$$

$$\lim_{x\to (e^{-1})^{+}} \frac{\log x}{\log x + 1}, \text{ limit for } x \to (e^{-1})^{+}$$

$$\lim_{x\to (e^{-1})^{-}} \frac{\log x}{\log x + 1}, \text{ limit for } x \to (e^{-1})^{-}$$

Considering the limits:

$$\lim_{x \to 0+} \frac{\log x}{\log x + 1}, \ \lim_{x \to +\infty} \frac{\log x}{\log x + 1}$$

Both limits have an indeterminate form:

$$\lim_{x \to 0+} \frac{\log x}{\log x + 1} = \frac{-\infty}{-\infty}$$

$$\lim_{x \to +\infty} \frac{\log x}{\log x + 1} = \frac{+\infty}{+\infty}$$

We can write:

$$\frac{\log x}{\log x + 1} = \frac{\log x}{\log x \left(1 + \frac{1}{\log x}\right)} = \frac{1}{1 + \frac{1}{\log x}}$$

So we get:

$$\lim_{x \to 0+} \frac{1}{\log x} = \frac{1}{-\infty} = 0$$

$$\lim_{x \to +\infty} \frac{1}{\log x} = \frac{1}{+\infty} = 0$$

Thus:

$$\lim_{x \to 0+} \frac{\log x}{\log x + 1} = \lim_{x \to 0+} \frac{1}{1 + \frac{1}{\log x}} = \frac{1}{1+0} = 1$$

$$\lim_{x \to +\infty} \frac{\log x}{\log x + 1} = \lim_{x \to +\infty} \frac{1}{1 + \frac{1}{\log x}} = \frac{1}{1+0} = 1$$

Let us now consider the limits:

$$\lim_{x\to(e^{-1})^+}\frac{\log x}{\log x+1} \text{ limit for } x\to\left(e^{-1}\right)^+$$

$$\lim_{x\to(e^{-1})^-}\frac{\log x}{\log x+1} \text{ limit for } x\to\left(e^{-1}\right)^-$$

In particular:

$$\lim_{x\to(e^{-1})^+}\log x=-1,\ \text{for } x>e^{-1}\text{: } \log x>-1 \;\rightarrow\; \log x+1>0$$

$$\lim_{x\to(e^{-1})^-}\log x=-1,\ \text{for } x<e^{-1}\text{: } \log x<-1 \;\rightarrow\; \log x+1<0$$

for which:

$$\lim_{x\to(e^{-1})^+}\frac{\log x}{\log x+1}=\frac{-1}{0^+}=-\infty$$
$$\lim_{x\to(e^{-1})^-}\frac{\log x}{\log x+1}=\frac{-1}{0^-}=+\infty$$

The graph of the function is shown in Figure 9.2:

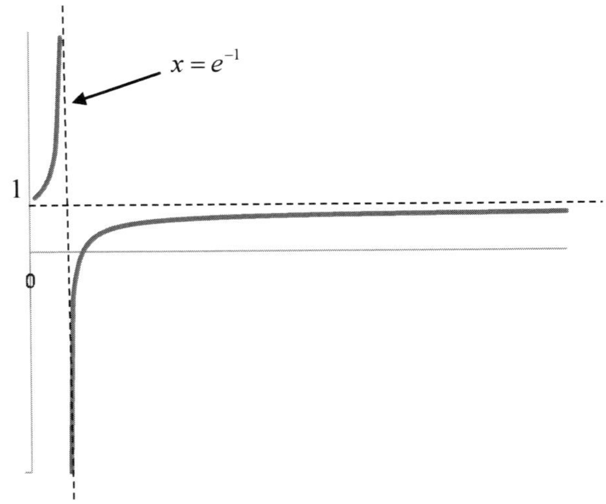

Figure 9.2. Graphical representation of the function of Eq. 9.3.

9.2.4. Problem 9.4.

Let:

$$f(x) = \frac{\log(x+1)}{x}$$

The domain of the function in the numerator is determined by the condition that the argument of logarithm is > 0:

$$x+1>0 \quad \Rightarrow \quad x>-1 \quad \Rightarrow \quad D: \ ]-1,+\infty[$$

From this we must exclude the values for which the denominator is equal to 0 (it must be $x \neq 0$).

The function is thus defined by:

$$f: A \to \boldsymbol{R} \qquad A =]-1,0[\cup]0,+\infty[=]-1,0[\cup \boldsymbol{R}^+$$

We want to calculate:

$$\lim_{x\to 0} \frac{\log(x+1)}{x}$$

the limit appears in the indeterminate form $\frac{0}{0}$, for $x \to 0^-$ and $x \to 0^+$:

$$\lim_{x\to 0} \frac{\log(x+1)}{x} = \frac{\log(0+1)}{0} = \frac{0}{0}$$

Considering one of the properties of logarithms (Eq. 4.37), we have:

$$\frac{\log(x+1)}{x} = \frac{1}{x} \cdot \log(x+1) = \log\left[(x+1)^{\frac{1}{x}}\right]$$

We perform a variable substitution:

$$z = \frac{1}{x} \qquad \Rightarrow \qquad x = \frac{1}{z}$$

So, when:

$$x \to 0 \qquad \Rightarrow \qquad z \to \pm\infty$$

the limit becomes:

$$\lim_{x\to 0} \frac{\log(x+1)}{x} = \lim_{x\to 0} \log\left[(x+1)^{\frac{1}{x}}\right] = \lim_{z\to +\infty} \log\left[\left(1+\frac{1}{z}\right)^{z}\right]$$

Where, recalling the definition of the Napier's number (Eq. 4.39a):

$$\lim_{z\to +\infty}\left(1+\frac{1}{z}\right)^{z} = e$$

and:

$$\lim_{z\to +\infty} \log\left[\left(1+\frac{1}{z}\right)^{z}\right] = \log e = 1$$

thus:

$$\lim_{x\to 0} \frac{\log(x+1)}{x} = 1$$

9.2.5. Problem 9.5

Let:

$$f(x) = \frac{e^{x}-1}{x}$$

The domain of the numerator is ***R***, so the function is defined by:

$$f : \boldsymbol{R} - \{0\} \rightarrow \boldsymbol{R}$$

We want to calculate:

$$\lim_{x \to 0} \frac{e^x - 1}{x}$$

the limit appears in the indeterminate form $\frac{0}{0}$, for $x \to 0^-$ and $x \to 0^+$:

$$\lim_{x \to 0} \frac{e^x - 1}{x} = \frac{e^0 - 1}{0} = \frac{0}{0}$$

We perform a variable substitution:

$$z = e^x - 1 \qquad \Rightarrow \qquad e^x = z + 1 \qquad \Rightarrow \qquad x = \log e^x = \log(z+1)$$

So, when:

$$x \to 0 \qquad \Rightarrow \qquad z \to 0$$

the limit becomes:

$$\lim_{x \to 0} \frac{e^x - 1}{x} = \lim_{z \to 0} \frac{z}{\log(z+1)} = \lim_{z \to 0} \frac{1}{\frac{\log(z+1)}{z}} = 1$$

As seen in the previous problem 9.4:

$$\lim_{z \to 0} \frac{\log(z+1)}{z} = 1$$

It is therefore:

$$\lim_{x \to 0} \frac{e^x - 1}{x} = 1$$

9.2.6. Problem 9.6

Let:

$$f(x) = \frac{\sin x}{\sin x + 1}$$

The domains of the numerator and denominator are equal to $\boldsymbol{R}$, from which it is necessary to remove the values in which the denominator is equal to zero:

$$\sin x + 1 = 0 \to \sin x = -1 \to x = \frac{3}{2}\pi + 2k\pi \equiv \left(\frac{3}{2} + 2k\right)\pi; \qquad k \in \boldsymbol{I}$$

The function is defined by:

$$f : \boldsymbol{R} - \left\{\left(\frac{3}{2} + 2k\right)\pi : k \in \boldsymbol{I}\right\} \to \boldsymbol{R}$$

In fact, for values:

$$x = \left(\frac{3}{2} + 2k\right)\pi; k \in \boldsymbol{I} \Rightarrow$$

$$\sin\left(\frac{3}{2} + 2k\right)\pi = \sin\left(\frac{3}{2}\pi + 2k\pi\right) = \sin\frac{3}{2}\pi = -1$$

the denominator equal to 0 and the function diverges.

In general, for $k \in \boldsymbol{I}$:

$$\sin 0 = \sin\pi = \sin\kappa\pi = 0$$

$$\sin\frac{\pi}{2} = \sin\left(\frac{\pi}{2} + 2\kappa\pi\right) = \sin\left(\frac{1}{2} + 2\kappa\right)\pi = 1$$

$$\sin\frac{3}{2}\pi = \sin\left(\frac{3}{2}\pi + 2\kappa\pi\right) = \sin\left(\frac{3}{2} + 2\kappa\right)\pi = -1$$

We can consider the limits:

$$\lim_{x\to k\pi}\frac{\sin x}{\sin x+1}$$

$$\lim_{x\to\left(\frac{1}{2}+2k\right)\pi}\frac{\sin x}{\sin x+1}$$

$$\lim_{x\to\left(\frac{3}{2}+2k\right)\pi}\frac{\sin x}{\sin x+1}$$

For the first two limits it holds:

$$\lim_{x\to k\pi}\frac{\sin x}{\sin x+1}=\frac{0}{0+1}=0$$

$$\lim_{x\to\left(\frac{1}{2}+2k\right)\pi}\frac{\sin x}{\sin x+1}=\frac{1}{1+1}=\frac{1}{2}$$

For the limit:

$$\lim_{x\to\left(\frac{3}{2}+2k\right)\pi}\frac{\sin x}{\sin x+1}$$

it can be observed that the denominator always tends to 0 from positive values:

$$|\sin x|\le 1 \quad\Rightarrow\quad \sin x\ge -1$$

For which (the numerator is negative):

$$\lim_{x\to\left(\frac{3}{2}+2k\right)\pi}\frac{\sin x}{\sin x+1}=\frac{-1}{0^+}=-\infty$$

The graph of the function is shown in Figure 9.3.

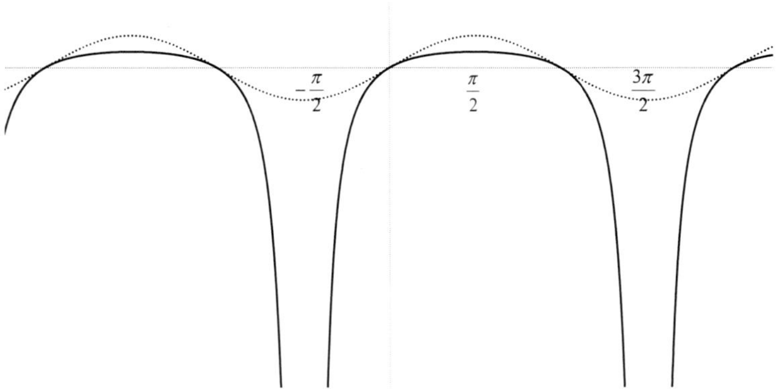

Figure 9.3. Graph of the function of Eq. 9.6 (the sine function is in dashed line).

9.2.7. Problem 9.7

Let:

$$f(x)=\frac{\sin x}{x}$$

We remove from R the values for which the denominator is equal to 0. Therefore the function is defined by:

$$f:\boldsymbol{R}-\{0\}\to\boldsymbol{R}$$

We want to calculate:

$$\lim_{x\to 0}\frac{\sin x}{x}$$

We have:

$$\lim_{x\to 0^+}\frac{\sin x}{x}=\frac{0^+}{0^+}$$

$$\lim_{x\to 0^-}\frac{\sin x}{x}=\frac{0^-}{0^-}$$

Both limits have an indeterminate form $\frac{0}{0}$.

By recalling one of the properties of the limits (Eq. 4.16e):

$$A \subseteq \boldsymbol{R}, \qquad x_0 \in D(A), \qquad f, g : A \to \boldsymbol{R}$$
$$\lim_{x \to x_0} f(x) = \alpha, \lim_{x \to x_0} g(x) = \beta, \qquad \alpha, \beta \in \boldsymbol{R}$$
$$h : A \to \boldsymbol{R}, f(x) \le h(x) \le g(x), \forall x \in A,$$
$$\alpha = \beta \Rightarrow \lim_{x \to x_0} h(x) = \alpha = \beta$$

in the present case it can be considered:

$$h(x) = \frac{\sin x}{x}, \qquad A = \boldsymbol{R} - \{0\}$$

If we find two functions respectively $\le$ and $\ge$ of the function $h(x)$, which have the same limit, then it is possible to calculate the limit of the function $h(x)$.

We can consider an arc $\widehat{AP}$ on the unit circle, see Figure 9.4.

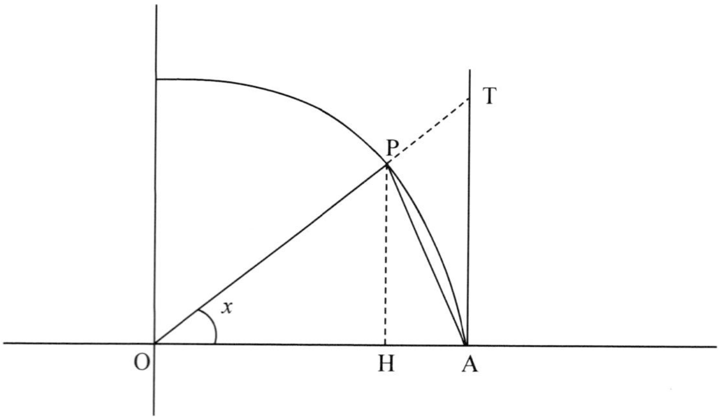

Figure 9.4. Representation of an arc on the unit circle.

In the graphical representation, we can identify the triangles $O\hat{P}A$ and $O\hat{T}A$ and the circular sector $O\widehat{A}\widehat{P}$ centered in O and delimited by the arc $\widehat{AP}$ of length $l\left(\widehat{AP}\right)$.

Considering the areas of these geometric figures, we observe that (the equality is valid only in the case $x = 0$):

$$Area\left(O\hat{P}A\right) \le Area\left(O\widehat{A}\widehat{P}\right) \le Area\left(O\hat{T}A\right)$$

If we explicitly calculate the areas, we have:

$$\frac{1}{2}OA \cdot PH \le \frac{1}{2}OA \cdot l\left(\widehat{A}\widehat{P}\right) \le \frac{1}{2}OA \cdot TA$$

that is, by simplifying $\frac{1}{2}OA$:

$$PH \le l\left(\widehat{A}\widehat{P}\right) \le TA$$

From the definition of angle in radians and from the definition of circular functions (Eq. 4.51 and 4.60) we have:

$$x = l\left(\widehat{A}\widehat{P}\right)$$

$$PH = \sin x$$

$$OH = \cos x$$

$$\frac{PH}{OH} = \frac{\sin x}{\cos x} = \tan x$$

The two triangles $O\hat{P}H$ and $O\hat{T}A$ are similar triangles, so for the similarity properties of the triangles:

$$\tan x = \frac{PH}{OH} = \frac{TA}{OA} = TA$$

where $OA = 1$ is the radius of the unit circle. It is therefore:

$$\sin x \le x \le \tan x$$

i.e.:

$$1 \le \frac{x}{\sin x} \le \tan x \cdot \frac{1}{\sin x}$$

where:

$$\tan x \cdot \frac{1}{\sin x} = \frac{\sin x}{\cos x} \cdot \frac{1}{\sin x} = \frac{1}{\cos x}$$

By substituting we get:

$$1 \le \frac{x}{\sin x} \le \frac{1}{\cos x}$$

By considering the reciprocal of the functions and consequently inverting the sign of inequalities:

$$1 \ge \frac{\sin x}{x} \ge \cos x$$

For the cosine it holds:

$$\lim_{x \to 0} \cos x = \cos 0 = 1$$

for which:

$$1 \le \lim_{x \to 0} \frac{\sin x}{x} \le 1$$

$$\lim_{x \to 0} \frac{\sin x}{x} = 1$$

Now we want to calculate the limits:

$$\lim_{x \to +\infty} \frac{\sin x}{x}$$

$$\lim_{x \to -\infty} \frac{\sin x}{x}$$

The sine function is a bounded function:

$$\sin x \in [-1,1] \qquad \forall x \in \boldsymbol{R}$$

which can also be written as:

$$|\sin x| \leq 1 \qquad \forall x \in \boldsymbol{R}$$

The numerator will never diverge for $x \to +\infty$, but it will only assume real values ≥ −1 and ≤ 1, thus:

$$\lim_{x \to +\infty} \frac{\sin x}{x} = 0$$

$$\lim_{x \to -\infty} \frac{\sin x}{x} = 0$$

9.2.8. Problem 9.8

Let:

$$f(x) = \frac{\log(3x+1)}{x}$$

Calculate:

$$\lim_{x \to 0} \frac{\log(3x+1)}{x}$$

The domain of definition of the function consists of the function domain in the numerator minus the values in which the denominator is equal to zero.

The domain of the numerator consists of the values for which the argument of the logarithm function is > 0:

$$\log : \boldsymbol{R}^+ \to \boldsymbol{R}$$

This is true when:

$$3x+1>0 \quad \Rightarrow \quad 3x>-1 \quad \Rightarrow \quad x>-\frac{1}{3}$$

The function is therefore defined by:

$$f:\left]-\frac{1}{3},+\infty\right[-\{0\}\rightarrow \boldsymbol{R} \quad \equiv \quad f:\left]-\frac{1}{3},0\right[\cup\left]0,+\infty\right[\rightarrow \boldsymbol{R}$$

We want to calculate:

$$\lim_{x\to 0}\frac{\log(3x+1)}{x}$$

the limit appears in the indeterminate form $\frac{0}{0}$.

We perform a variable substitution:

$$z = 3x \qquad \rightarrow \qquad x=\frac{z}{3}$$

So, when:

$$x\rightarrow 0 \qquad \rightarrow \qquad z\rightarrow 0$$

The limit becomes:

$$\lim_{x\to 0}\frac{\log(3x+1)}{x}=\lim_{z\to 0}\frac{\log(z+1)}{\frac{z}{3}}=3\cdot\lim_{z\to 0}\frac{\log(z+1)}{z}$$

As seen in the problem 9.4:

$$\lim_{x\to 0}\frac{\log(x+1)}{x}=1$$

Thus:

$$\lim_{x\to 0}\frac{\log(3x+1)}{x}=3\cdot\lim_{z\to 0}\frac{\log(z+1)}{z}=3$$

9.2.9. Problem 9.9

Let:

$$f(x)=e^{-3x^2+2}$$

Calculate:

$$\lim_{x\to\pm\infty} e^{-3x^2+2}$$

In this case the function definition domain is ***R***:

$$f:\boldsymbol{R}\to\boldsymbol{R}$$

The limit can be written as:

$$\lim_{x\to\pm\infty} e^{-3x^2+2}=\lim_{x\to\pm\infty} e^{-3x^2}\cdot e^2=e^2\cdot\lim_{x\to\pm\infty} e^{-3x^2}$$

It holds:

$$\lim_{x\to+\infty} e^{-x}=e^{-\infty}=0$$

In the present case, the exponent x is in turn raised to the square:

$$\lim_{x\to+\infty} e^{-3x^2}=\lim_{x\to-\infty} e^{-3x^2}=e^{-\infty}=0$$

Thus:

$$\lim_{x\to\pm\infty} e^{-3x^2+2}=0$$

9.2.10. Problem 9.10

Let:

$$f(x) = \frac{\log x}{x-1}$$

Calculate:

$$\lim_{x \to 1} \frac{\log x}{x-1}$$

The domain of definition of the function consists of the function domain in the numerator, in this case $\boldsymbol{R}^{+}$, minus the values in which the denominator is equal to zero:

$$x - 1 = 0 \quad \Rightarrow \quad x = 1$$

The function is therefore defined by:

$$f : \boldsymbol{R}^{+} - \{1\} \to \boldsymbol{R} \quad \equiv \quad f :]0,1[\cup]1,+\infty[\to \boldsymbol{R}$$

We want to calculate:

$$\lim_{x \to 1} \frac{\log x}{x-1}$$

The limit appears in the indeterminate form $\frac{0}{0}$.

We perform a variable substitution:

$$z = x - 1 \quad \to \quad x = z + 1$$

So, when:

$$x \to 1 \quad \to \quad z \to 0$$

The limit becomes (see problem 9.4):

$$\lim_{x \to 1} \frac{\log x}{x-1} = \lim_{z \to 0} \frac{\log(z+1)}{z} = 1$$

9.2.11. Problem 9.11

Let:

$$f(x) = e^{4x^3 - 2x + 1}$$

Calculate:

$$\lim_{x \to +\infty} e^{4x^3 - 2x + 1}$$

In this case the function definition domain is ***R***:

$$f : \boldsymbol{R} \to \boldsymbol{R}$$

The limit appears in an indeterminate form:

$$\lim_{x \to +\infty} e^{4x^3 - 2x + 1} = \lim_{x \to +\infty} e^{4x^3} \cdot e^{-2x+1} = e^{+\infty} \cdot e^{-\infty} = +\infty \cdot 0$$

We perform a variable substitution:

$$z = 4x^3 - 2x + 1 = 2x\left(2x^2 - 1 + \frac{1}{2x}\right)$$

So, when:

$$x \to +\infty \quad \Rightarrow \quad z \to (+\infty) \cdot (+\infty - 1 + 0) = (+\infty) \cdot (+\infty) = +\infty$$

the limit becomes:

$$\lim_{x \to +\infty} e^{4x^3 - 2x + 1} = \lim_{z \to +\infty} e^{z} = +\infty$$

An alternative calculation method, without variable substitution, it is the following:

$$\lim_{x \to +\infty} e^{4x^3-2x+1} = \lim_{x \to +\infty} e^{4x^3-2x} \cdot e^{+1} =$$

$$= \lim_{x \to +\infty} e \cdot e^{2x\left(2x^2-1\right)} = e \cdot e^{+\infty(+\infty)} = e^{+\infty} = +\infty$$

9.2.12. Problem 9.12

Let:

$$f(x) = e^{4x^3-2x+1}$$

Calculate:

$$\lim_{x \to -\infty} e^{4x^3-2x+1}$$

In this case the function definition domain is ***R***:

$$f : \boldsymbol{R} \to \boldsymbol{R}$$

The limit appears in an indeterminate form:

$$\lim_{x \to -\infty} e^{4x^3-2x+1} = \lim_{x \to -\infty} e^{4x^3} \cdot e^{-2x+1} = e^{-\infty} \cdot e^{+\infty} = 0 \cdot (+\infty)$$

We perform a variable substitution:

$$z = 4x^3 - 2x + 1 = 2x\left(2x^2 - 1 + \frac{1}{2x}\right)$$

So, when:

$$x \to -\infty \quad \Rightarrow \quad z \to (-\infty)\cdot(+\infty - 1 + 0) = (-\infty)\cdot(+\infty) = -\infty$$

the limit becomes:

$$\lim_{x \to -\infty} e^{4x^3-2x+1} = \lim_{z \to -\infty} e^{z} = 0$$

An alternative calculation method, without variable substitution, it is the following:

$$\lim_{x\to-\infty} e^{4x^3-2x+1} = \lim_{x\to-\infty} e^{4x^3-2x}\cdot e^{+1} =$$
$$= \lim_{x\to-\infty} e\cdot e^{2x\left(2x^2-1\right)} = e\cdot e^{-\infty(+\infty)} = e^{-\infty} = 0$$

9.2.13. Problem 9.13

Let:

$$f(x) = \frac{\sin 2x}{x}$$

Calculate:

$$\lim_{x\to 0}\frac{\sin 2x}{x}$$

The domain of definition of the function consists of the function domain in the numerator, in this case ***R***, minus the values in which the denominator is equal to zero:

$$x = 0$$

The function is therefore defined by:

$$f : \boldsymbol{R} - \{0\} \to \boldsymbol{R}$$

The limit appears in the indeterminate form $\frac{0}{0}$:

$$\lim_{x\to 0}\frac{\sin 2x}{x} = \frac{0}{0}$$

We perform a variable substitution:

$$z = 2x \qquad \to \qquad x = \frac{z}{2}$$

So, when:

$$x \to 0 \quad \Rightarrow \quad z \to 0$$

the limit becomes:

$$\lim_{x\to 0}\frac{\sin 2x}{x} = \lim_{z\to 0}\frac{\sin z}{\frac{z}{2}} = \lim_{z\to 0} 2\cdot\frac{\sin z}{z} = 2\cdot\lim_{z\to 0}\frac{\sin z}{z}$$

As seen in the problem 9.7:

$$\lim_{z\to 0}\frac{\sin z}{z} = 1$$

thus:

$$\lim_{x\to 0}\frac{\sin 2x}{x} = 2$$

9.2.14. Problem 9.14

Let:

$$f(x) = \frac{x - \sin x}{x + \sin x}$$

Calculate:

$$\lim_{x\to 0}\frac{x - \sin x}{x + \sin x}$$

The domain of definition of the function is given by the intersection of the domains of the numerator and denominator, in this case ***R***, minus the values in which the denominator is equal to 0:

$$x + \sin x = 0 \quad \Rightarrow \quad x = 0$$

$x = 0$ it is the only value for which the denominator is $= 0$.
The function is therefore defined by:

$$f : \boldsymbol{R} - \{0\} \to \boldsymbol{R}$$

The limit appears in the indeterminate form $\frac{0}{0}$:

$$\lim_{x\to 0}\frac{x-\sin x}{x+\sin x}=\frac{0}{0}$$

By changing the form of the function, we get:

$$\lim_{x\to 0}\frac{x-\sin x}{x+\sin x}=\lim_{x\to 0}\frac{x\left(1-\dfrac{\sin x}{x}\right)}{x\left(1+\dfrac{\sin x}{x}\right)}=\lim_{x\to 0}\frac{1-\dfrac{\sin x}{x}}{1+\dfrac{\sin x}{x}}$$

It holds (see problem 9.7):

$$\lim_{x\to 0}\frac{\sin x}{x}=1$$

thus:

$$\lim_{x\to 0}\frac{x-\sin x}{x+\sin x}=\frac{1-1}{1+1}=\frac{0}{2}=0$$

9.2.15. Problem 9.15

Let:

$$f(x)=\frac{x-\sin x}{x+\sin x}$$

Calculate:

$$\lim_{x\to+\infty}\frac{x-\sin x}{x+\sin x}$$

The domain of definition of the function is given by the intersection of the domains of the numerator and denominator, in this case ***R***, minus the values in which the denominator is equal to 0:

$$x+\sin x=0 \quad \Rightarrow \quad x=0$$

$x=0$ it is the only value for which the denominator is $=0$.
The function is therefore defined by:

$$f:\boldsymbol{R}-\{0\}\to\boldsymbol{R}$$

The sine function is a bounded function, so the limit appears in the indeterminate form $\frac{+\infty}{+\infty}$:

$$\lim_{x\to+\infty}\frac{x-\sin x}{x+\sin x}=\frac{+\infty}{+\infty}$$

In fact:

$$\sin x\in[-1,1],\forall x\in\boldsymbol{R}$$

or also:

$$|\sin x|\leq 1,\forall x\in\boldsymbol{R}$$

By changing the form of the function, we get:

$$\lim_{x\to+\infty}\frac{x-\sin x}{x+\sin x}=\lim_{x\to+\infty}\frac{x\left(1-\frac{\sin x}{x}\right)}{x\left(1+\frac{\sin x}{x}\right)}=\lim_{x\to+\infty}\frac{1-\frac{\sin x}{x}}{1+\frac{\sin x}{x}}$$

As seen previously (problem 9.7):

$$\lim_{x\to+\infty}\frac{\sin x}{x}=0$$

thus:

$$\lim_{x\to+\infty}\frac{x-\sin x}{x+\sin x}=\lim_{x\to+\infty}\frac{1-\frac{\sin x}{x}}{1+\frac{\sin x}{x}}=\frac{1-0}{1+0}=1$$

9.2.16. Problem 9.16

Let:

$$f(x)=\frac{\left(1-\sin^2 x\right)^2}{\cos x}$$

Calculate:

$$\lim_{x\to\frac{\pi}{2}}\frac{\left(1-\sin^2 x\right)^2}{\cos x}$$

The domain of definition of the function is given by the intersection of the domains of the numerator and denominator, in this case ***R***, minus the values in which the denominator is equal to 0:

$$\cos x=0 \quad\Rightarrow\quad x=\left(k+\frac{1}{2}\right)\pi;k\in I$$

The function is therefore defined by:

$$f:\boldsymbol{R}-\left\{\left(k+\frac{1}{2}\right)\pi:k\in I\right\}\to\boldsymbol{R}$$

The limit appears in the indeterminate form $\frac{0}{0}$:

$$\lim_{x\to\frac{\pi}{2}}\frac{\left(1-\sin^2 x\right)^2}{\cos x}=\frac{0}{0}$$

It holds (Eq. 4.54):

$$\sin^2 x+\cos^2 x=1 \quad\Rightarrow\quad \cos^2 x=1-\sin^2 x$$

from which:

$$\lim_{x\to\frac{\pi}{2}}\frac{\left(1-\sin^2 x\right)^2}{\cos x}=\lim_{x\to\frac{\pi}{2}}\frac{\left(\cos^2 x\right)^2}{\cos x}=\lim_{x\to\frac{\pi}{2}}\frac{\cos^4 x}{\cos x}=\lim_{x\to\frac{\pi}{2}}\cos^3 x$$

Thus:

$$\lim_{x\to\frac{\pi}{2}}\frac{\left(1-\sin^2 x\right)^2}{\cos x}=\lim_{x\to\frac{\pi}{2}}\cos^3 x=0$$

9.2.17. Problem 9.17

Let:

$$f(x)=\frac{1-\cos x}{x^2}$$

Calculate:

$$\lim_{x\to 0}\frac{1-\cos x}{x^2}$$

The domain of definition of the function is given by the intersection of the domains of the numerator and denominator, in this case ***R***, minus the values in which the denominator is equal to 0:

$$x^2=0 \quad\Rightarrow\quad x=0$$

The function is therefore defined by:

$$f : \boldsymbol{R} - \{0\} \to \boldsymbol{R}$$

The limit appears in the indeterminate form $\frac{0}{0}$:

$$\lim_{x\to 0}\frac{1-\cos x}{x^2} = \frac{1-1}{0} = \frac{0}{0}$$

We can apply De L'Hôpital's rule (Eq. 5.35 and 5.37):

$$\lim_{x\to 0}\frac{f(x)}{g(x)} = \lim_{x\to 0}\frac{f'(x)}{g'(x)}$$

$$f'(x) = \frac{d}{dx}(1-\cos x) = \sin x$$

$$g'(x) = \frac{d}{dx}\left(x^2\right) = 2x$$

So, see also problem 9.7:

$$\lim_{x\to 0}\frac{f'(x)}{g'(x)} = \lim_{x\to 0}\frac{\sin x}{2x} = \frac{1}{2}\cdot\lim_{x\to 0}\frac{\sin x}{x} = \frac{1}{2}\cdot 1 = \frac{1}{2}$$

The limit is therefore:

$$\lim_{x\to 0}\frac{1-\cos x}{x^2} = \frac{1}{2}$$

Chapter 10

Calculation of Function Derivatives

10.1. Problems

Indicate the domain of definition of the following functions and calculate the derivative.

$$f(x)=\frac{x^4}{4}-\frac{x^3}{3}-x-2 \tag{10.1}$$

$$f(x)=e^{x^2+x} \tag{10.2}$$

$$f(x)=x\cdot\log x \tag{10.3}$$

$$f(x)=\sqrt[3]{(1+x)^2} \tag{10.4}$$

$$f(x)=\log(\log x) \tag{10.5}$$

$$f(x)=\frac{\log x-1}{\log x+1} \tag{10.6}$$

$$f(x)=\sin x+\cos x+\tan x \tag{10.7}$$

$$f(x)=\frac{1+\sin x}{1-\sin x} \tag{10.8}$$

$$f(x) = \tan\left(-\frac{x}{3}+2\right) \tag{10.9}$$

$$f(x) = \sqrt[3]{\cos^2 x} \tag{10.10}$$

$$f(x) = \sin(\log x) \tag{10.11}$$

$$f(x) = \frac{\log x + 1}{\log x - 1} \tag{10.12}$$

$$f(x) = \log(\sin x) \tag{10.13}$$

$$f(x) = \log\sqrt{1+x^2} \tag{10.14}$$

$$f(x) = e^{\sin x} \tag{10.15}$$

$$f(x) = \frac{1}{\sin x + 1} \tag{10.16}$$

$$f(x) = \frac{\sin x}{\sin x + 1} \tag{10.17}$$

$$f(x) = \sqrt{1+x^a} \tag{10.18}$$

10.2. SOLUTIONS

10.2.1. Problem 10.1

Calculate the derivative of the function:

$$f(x) = \frac{x^4}{4} - \frac{x^3}{3} - x - 2$$

The domain of definition of the function is ***R***:

$f: \boldsymbol{R} \rightarrow \boldsymbol{R}$

The function is a sum of functions, so the derivative of a sum is equal to the sum of the derivatives (Eq. 5.11a):

$$(f+g)'(x) = f'(x) + g'(x)$$

$$\frac{d}{dx}\left(\frac{x^4}{4} - \frac{x^3}{3} - x - 2\right) = 4 \cdot \frac{x^3}{4} - 3 \cdot \frac{x^2}{3} - 1 = x^3 - x^2 - 1$$

10.2.2. Problem 10.2

Calculate the derivative of the function:

$f(x) = e^{x^2+x}$

The domain of definition of the function is $\boldsymbol{R}$:

$f: \boldsymbol{R} \rightarrow \boldsymbol{R}$

The function is a composite function, so the derivative is (Eq. 5.11e):

$$(g(f))'(x) = g'(y) \cdot f'(x) = g'(f(x)) \cdot f'(x)$$

$$\frac{d}{dx}\left(e^{x^2+x}\right) = e^{x^2+x} \cdot \frac{d}{dx}\left(x^2 + x\right) = (2x+1) \cdot e^{x^2+x}$$

10.2.3. Problem 10.3

Calculate the derivative of the function:

$f(x) = x \cdot \log x$

The domain of the function coincides with the domain of the logarithm function:

$f: \boldsymbol{R}^{+} \rightarrow \boldsymbol{R}$

The function is a product of functions, so the derivative is (Eq. 5.11b):

$$(f \cdot g)'(x) = f'(x)g(x) + f(x)g'(x)$$

$$\frac{d}{dx}(x \cdot \log x) = 1 \cdot \log x + x \cdot \frac{1}{x} = 1 + \log x$$

10.2.4. Problem 10.4

Calculate the derivative of the function:

$$f(x) = \sqrt[3]{(1+x)^2}$$

The domain of definition of the function is $\boldsymbol{R}$:

$$f : \boldsymbol{R} \to \boldsymbol{R}$$

The derivative is:

$$\frac{d}{dx}\left(\sqrt[3]{(1+x)^2}\right) = \frac{d}{dx}\left[(1+x)^2\right]^{\frac{1}{3}} = \frac{d}{dx}(1+x)^{\frac{2}{3}} = \frac{2}{3}(1+x)^{\frac{2}{3}-1} =$$

$$= \frac{2}{3}(1+x)^{-\frac{1}{3}} = \frac{2}{3} \cdot \frac{1}{(1+x)^{\frac{1}{3}}} = \frac{2}{3 \cdot \sqrt[3]{1+x}}$$

10.2.5. Problem 10.5

Calculate the derivative of the function:

$$f(x) = \log(\log x)$$

The function is a composite function of the logarithm of a logarithm. The domain of the function coincides with the values for which the argument is:

$$\log x > 0 \quad \Rightarrow \quad x > 1 \quad \Rightarrow \quad D :]1, +\infty[$$

The function is therefore defined by:

$$f :]1,+\infty[\to \boldsymbol{R}$$

The function is a composite function, so the derivative is (Eq. 5.11e):

$$\frac{d}{dx}[\log(\log x)] = \frac{1}{\log x} \cdot \frac{d}{dx}(\log x) = \frac{1}{\log x} \cdot \frac{1}{x} = \frac{1}{x \cdot \log x}$$

10.2.6. Problem 10.6

Calculate the derivative of the function:

$$f(x) = \frac{\log x - 1}{\log x + 1}$$

The domain of definition of the function consists of the logarithm domain (found both in the numerator and in the denominator), in this case $\boldsymbol{R}^+$, minus the values in which the denominator is equal to 0:

$$\log x + 1 = 0 \Rightarrow \log x = -1 \Rightarrow e^{\log x} = e^{-1} \Rightarrow x = e^{-1}$$

The domain of the function is:

$$D = \boldsymbol{R}^+ - \{e^{-1}\} \equiv]0, e^{-1}[\cup]e^{-1}, +\infty[$$

The function is defined by:

$$f : \boldsymbol{R}^+ - \{e^{-1}\} \to \boldsymbol{R}$$

The function is a ratio between two functions, so the derivative is (Eq. 5.11d):

$$\left(\frac{f}{g}\right)'(x) = \frac{f'(x)g(x) - f(x)g'(x)}{g^2(x)}$$

$$\frac{d}{dx}\left(\frac{\log x-1}{\log x+1}\right)=\frac{\frac{1}{x}\cdot(\log x+1)-(\log x-1)\cdot\frac{1}{x}}{(\log x+1)^2}=$$

$$=\frac{\log x+1-\log x+1}{x\cdot(\log x+1)^2}=\frac{2}{x\cdot(\log x+1)^2}$$

Thus:

$$\frac{d}{dx}\left(\frac{\log x-1}{\log x+1}\right)=\frac{2}{x\cdot(\log x+1)^2}$$

10.2.7. Problem 10.7

Calculate the derivative of the function:

$$f(x)=\sin x+\cos x+\tan x$$

The domain of definition of the function coincides with the tangent function domain (for sine and cosine functions the domain is $\boldsymbol{R}$):

$$f:\boldsymbol{R}-\left\{\left(k+\frac{1}{2}\right)\pi : k\in I\right\}\to\boldsymbol{R}$$

The function is a sum of functions, so the derivative is:

$$\frac{d}{dx}(\sin x+\cos x+\tan x)=\cos x-\sin x+\frac{1}{\cos^2 x}=$$

$$=\cos x-\sin x+\frac{\sin^2 x+\cos^2 x}{\cos^2 x}=\cos x-\sin x+\frac{\sin^2 x}{\cos^2 x}+\frac{\cos^2 x}{\cos^2 x}=$$

$$=\cos x-\sin x+1+\tan^2 x$$

Thus:

$$\frac{d}{dx}(\sin x+\cos x+\tan x)=\cos x-\sin x+1+\tan^2 x$$

10.2.8. Problem 10.8

Calculate the derivative of the function:

$$f(x)=\frac{1+\sin x}{1-\sin x}$$

The domain of definition of the function consists of the sine function domain (both in the numerator and denominator), therefore ***R***, minus the values in which the denominator is equal to 0:

$$1-\sin x=0 \Rightarrow \sin x=1 \Rightarrow x=\left(2k+\frac{1}{2}\right)\pi;k\in I$$

The function is therefore defined by:

$$f:\boldsymbol{R}-\left\{\left(2k+\frac{1}{2}\right)\pi:k\in I\right\}\rightarrow \boldsymbol{R}$$

The function is a ratio between two functions, so the derivative is (Eq. 5.11d):

$$\frac{d}{dx}\left(\frac{1+\sin x}{1-\sin x}\right)=\frac{\cos x(1-\sin x)-(1+\sin x)(-\cos x)}{(1-\sin x)^2}=$$

$$=\frac{\cos x-\sin x\cos x+\cos x+\sin x\cos x}{(1-\sin x)^2}=\frac{2\cos x}{(1-\sin x)^2}$$

Thus:

$$\frac{d}{dx}\left(\frac{1+\sin x}{1-\sin x}\right)=\frac{2\cos x}{(1-\sin x)^2}$$

10.2.9. Problem 10.9

Calculate the derivative of the function:

$$f(x) = \tan\left(-\frac{x}{3}+2\right)$$

The domain of the tangent function would be:

$$\tan : \boldsymbol{R} - \left\{\left(k+\frac{1}{2}\right)\pi : k \in I\right\} \rightarrow \boldsymbol{R}$$

In this case, considering the argument of the tangent function, it must be:

$$-\frac{x}{3}+2 \neq \left(k+\frac{1}{2}\right)\pi; k \in I \Rightarrow \frac{x}{3} \neq 2-\left(k+\frac{1}{2}\right)\pi \Rightarrow x \neq 6-3\left(k+\frac{1}{2}\right)\pi$$

The domain of the function is:

$$D = \boldsymbol{R} - \left\{6-3\left(k+\frac{1}{2}\right)\pi : k \in I\right\}$$

The function is therefore defined by:

$$f : \boldsymbol{R} - \left\{6-3\left(k+\frac{1}{2}\right)\pi : k \in I\right\} \rightarrow \boldsymbol{R}$$

The derivative is:

$$\frac{d}{dx}\left[\tan\left(-\frac{x}{3}+2\right)\right] = \frac{1}{\cos^2\left(-\frac{x}{3}+2\right)} \cdot \frac{d}{dx}\left(-\frac{x}{3}+2\right) =$$

$$= -\frac{1}{3 \cdot \cos^2\left(-\frac{x}{3}+2\right)}$$

Thus:

$$\frac{d}{dx}\left[\tan\left(-\frac{x}{3}+2\right)\right] = -\frac{1}{3 \cdot \cos^2\left(-\frac{x}{3}+2\right)}$$

10.2.10. Problem 10.10

Calculate the derivative of the function:

$$f(x)=\sqrt[3]{\cos^2 x}$$

The domain of definition of the function is $\boldsymbol{R}$:

$$f:\boldsymbol{R}\to\boldsymbol{R}$$

The derivative is:

$$\frac{d}{dx}\left(\sqrt[3]{\cos^2 x}\right)=\frac{d}{dx}(\cos x)^{\frac{2}{3}}=\frac{2}{3}\cdot(\cos x)^{\frac{2}{3}-1}\cdot(-\sin x)=$$
$$=-\frac{2}{3}\cdot\sin x\cdot(\cos x)^{-\frac{1}{3}}=-\frac{2\sin x}{3\cdot\sqrt[3]{\cos x}}$$

Thus:

$$\frac{d}{dx}\left(\sqrt[3]{\cos^2 x}\right)=-\frac{2\sin x}{3\cdot\sqrt[3]{\cos x}}$$

10.2.11. Problem 10.11

Calculate the derivative of the function:

$$f(x)=\sin(\log x)$$

The argument of the sine function is the logarithm function, for which it holds:

$$\sin:\boldsymbol{R}\to\boldsymbol{R}$$
$$\log:\boldsymbol{R}^{+}\to\boldsymbol{R}$$

The domain of definition of the function is $\boldsymbol{R}^{+}$:

$$f:\boldsymbol{R}^{+}\to\boldsymbol{R}$$

The derivative is:

$$\frac{d}{dx}[\sin(\log x)] = \cos(\log x)\cdot\frac{1}{x} = x^{-1}\cdot\cos(\log x)$$

10.2.12. Problem 10.12

Calculate the derivative of the function:

$$f(x) = \frac{\log x + 1}{\log x - 1}$$

The domain of definition of the function consists of the logarithm domain (found both in the numerator and in the denominator), in this case $\boldsymbol{R}^+$, minus the values in which the denominator is equal to 0:

$$\log x - 1 = 0 \Rightarrow \log x = 1 \quad \Rightarrow \quad e^{\log x} = e^1 \Rightarrow x = e$$

The domain of the function is:

$$D = \boldsymbol{R}^+ - \{e\} \equiv]0, e[\cup]e, +\infty[$$

The function is therefore defined by:

$$f : \boldsymbol{R}^+ - \{e\} \rightarrow \boldsymbol{R}$$

The function is a ratio between two functions, so the derivative is:

$$\left(\frac{f}{g}\right)'(x) = \frac{f'(x)g(x) - f(x)g'(x)}{g^2(x)}$$

$$\frac{d}{dx}\left(\frac{\log x + 1}{\log x - 1}\right) = \frac{\frac{1}{x}\cdot(\log x - 1) - (\log x + 1)\cdot\frac{1}{x}}{(\log x - 1)^2} =$$

$$= \frac{\log x - 1 - \log x - 1}{x\cdot(\log x - 1)^2} = \frac{-2}{x\cdot(\log x - 1)^2}$$

Thus:

$$\frac{d}{dx}\left(\frac{\log x+1}{\log x-1}\right)=-\frac{2}{x\cdot(\log x-1)^2}$$

10.2.13. Problem 10.13

Calculate the derivative of the function:

$$f(x)=\log(\sin x)$$

The logarithm function is defined by:

$$f(x)=\log x \qquad f:\boldsymbol{R}^+\to\boldsymbol{R} \quad \Rightarrow \quad x\in\boldsymbol{R}^+$$

In the present case, the argument of the logarithm function is the sine function, so the function can only be defined for the values for which it holds $\sin x > 0$.

The graph of the sine function is shown in Figure 10.1.

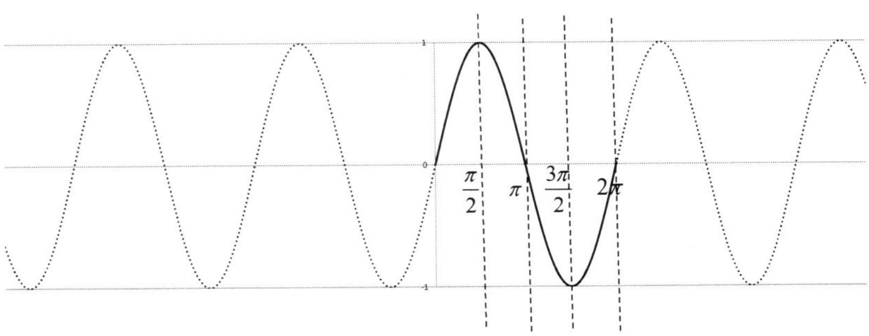

Figure 10.1. Graph of the sine function.

From the graph of the sine function (Figure 10.1) it can be observed that:

$$\sin x>0 \quad \Leftrightarrow \quad x\in\left]2k\pi,(2k+1)\pi\right[\quad ,k\in\boldsymbol{I}$$

The set of intervals thus defined is the domain of the function:

$$f(x)=\log(\sin x) \;\; f:\left]2k\pi,(2k+1)\pi\right[\to\boldsymbol{R} \quad ,k\in\boldsymbol{I}$$

For example:

$$k = 1 \quad \Rightarrow \quad]2k\pi, (2k+1)\pi[=]2\pi, 3\pi[$$
$$k = 0 \quad \Rightarrow \quad]2k\pi, (2k+1)\pi[=]0, \pi[$$
$$k = -1 \quad \Rightarrow \quad]2k\pi, (2k+1)\pi[=]-2\pi, -\pi[$$

The derivative is:

$$\frac{d}{dx}[\log(\sin x)] = \frac{1}{\sin x} \cdot \frac{d}{dx}(\sin x) = \frac{1}{\sin x} \cdot \cos x = \cot x$$

10.2.14. Problem 10.14

Calculate the derivative of the function:

$$f(x) = \log\sqrt{1+x^2}$$

The domain of definition of the function is $\boldsymbol{R}$.
In fact, the logarithm function is defined by:

$$f(x) = \log x \qquad f : \boldsymbol{R}^+ \to \boldsymbol{R} \quad \Rightarrow \quad x \in \boldsymbol{R}^+$$

In the present case:

$$f(x) = \log\sqrt{1+x^2} \quad \Rightarrow \quad \sqrt{1+x^2} \geq 1 \quad , \quad \forall x \in \boldsymbol{R}$$

The function is therefore defined by:

$$f : \boldsymbol{R} \to \boldsymbol{R}$$

f(x) is a composite function and its derivative is:

$$\frac{d}{dx}\left(\log\sqrt{1+x^2}\right) = \frac{d}{dx}\left(\log\left(1+x^2\right)^{\frac{1}{2}}\right) = \frac{1}{2}\frac{d}{dx}\left(\log\left(1+x^2\right)\right) =$$
$$= \frac{1}{2} \cdot \frac{1}{1+x^2} \cdot 2x = \frac{x}{1+x^2}$$

Thus:

$$\frac{d}{dx}\left(\log\sqrt{1+x^2}\right)=\frac{x}{1+x^2}$$

10.2.15. Problem 10.15

Calculate the derivative of the function:

$$f(x)=e^{\sin x}$$

The domain of definition of the function is $\boldsymbol{R}$:

$$f:\boldsymbol{R}\rightarrow\boldsymbol{R}$$

The derivative is:

$$\frac{d}{dx}\left(e^{\sin x}\right)=e^{\sin x}\cdot\cos x$$

10.2.16. Problem 10.16

Calculate the derivative of the function:

$$f(x)=\frac{1}{\sin x+1}$$

The domain of definition of the function is $\boldsymbol{R}$ minus the values in which the denominator is equal to zero:

$$\sin x+1=0 \quad\Rightarrow\quad \sin x=-1$$

This condition identifies the values:

$$\left\{\left(2k-\frac{1}{2}\right)\pi : k\in\boldsymbol{I}\right\}$$

The function is therefore defined by:

$$f:\boldsymbol{R}-\left\{\left(2k-\frac{1}{2}\right)\pi : k\in \boldsymbol{I}\right\}\to \boldsymbol{R}$$

The function is a reciprocal of a function, so the derivative is (Eq. 5.11c):

$$\frac{d}{dx}\left(\frac{1}{g(x)}\right)=-\frac{g'(x)}{g^2(x)}$$

$$\frac{d}{dx}\left(\frac{1}{\sin x+1}\right)=-\frac{\cos x}{(\sin x+1)^2}$$

10.2.17. Problem 10.17

Calculate the derivative of the function:

$$f(x)=\frac{\sin x}{\sin x+1}$$

The domain of definition of the function is $\boldsymbol{R}$ minus the values in which the denominator is equal to zero, so as seen in problem 10.16:

$$f:\boldsymbol{R}-\left\{\left(2k-\frac{1}{2}\right)\pi : k\in \boldsymbol{I}\right\}\to \boldsymbol{R}$$

The derivative is:

$$\frac{d}{dx}\left(\frac{\sin x}{\sin x+1}\right)=\frac{\cos x\cdot(\sin x+1)-\sin x\cdot\cos x}{(\sin x+1)^2}=$$
$$=\frac{\cos x\cdot\sin x+\cos x-\sin x\cdot\cos x}{(\sin x+1)^2}$$

Thus:

$$\frac{d}{dx}\left(\frac{\sin x}{\sin x+1}\right)=\frac{\cos x}{(\sin x+1)^2}$$

10.2.18. Problem 10.18

Calculate the derivative of the function:

$$f(x)=\sqrt{1+x^a}$$

The domain of definition of the function depends on the value of a.
If a is an even number, the function is defined by:

$$f:\boldsymbol{R}\rightarrow\boldsymbol{R},\ a=2k, k\in\boldsymbol{I}$$

If a is an odd number, the function is defined by:

$$f:[-1,+\infty[\rightarrow\boldsymbol{R}\quad a=2k+1, k\in\boldsymbol{I}$$

More generally, if a is a real number that does not fall in the previous cases (Eq. 4.47):

$$f:\boldsymbol{R}^+\rightarrow\boldsymbol{R}\qquad a\in\boldsymbol{R}$$

or, with alternative writing:

$$f:]0,+\infty[\rightarrow\boldsymbol{R}\qquad a\in\boldsymbol{R}$$

The derivative is:

$$\frac{d}{dx}\left(\sqrt{1+x^a}\right)=\frac{d}{dx}(1+x^a)^{\frac{1}{2}}=\frac{1}{2}\cdot(1+x^a)^{\frac{1}{2}-1}\cdot\frac{d}{dx}(x^a)=$$
$$=\frac{1}{2}\cdot(1+x^a)^{-\frac{1}{2}}\cdot ax^{a-1}$$

or, in equivalent form:

$$\frac{d}{dx}\left(\sqrt{1+x^a}\right)=\frac{ax^{a-1}}{2\left(1+x^a\right)^{\frac{1}{2}}}\equiv\frac{ax^{a-1}}{2\sqrt{1+x^a}}$$

Chapter 11

PROBLEMS RELATED TO THE STUDY OF FUNCTIONS

11.1. PROBLEMS

Study the following functions.

Specifically, indicate the domain of definition of the function and determine if they exist and what are the values of:

- Intersection with the axes.
- Asymptotes.
- Maximum and/or minimum points.
- Inflection points.

Draw the graph of the function.

$$f(x) = x^3 \tag{11.1}$$

$$f(x) = x^4 \tag{11.2}$$

$$f(x) = \frac{x^4}{4} - \frac{x^3}{3} \tag{11.3}$$

$$f(x) = \frac{x-1}{x^2} \tag{11.4}$$

$$f(x) = x^2 e^x \tag{11.5}$$

$$f(x) = x^2 \log x \tag{11.6}$$

$$f(x) = e^{-x^2} \tag{11.7}$$

$$f(x) = x - \sqrt{x^2 - 1} \tag{11.8}$$

$$f(x) = xe^{-x} \tag{11.9}$$

$$f(x) = (x-1)\log(x-1) \tag{11.10}$$

$$f(x) = \frac{\log x}{x^3} \tag{11.11}$$

$$f(x) = \frac{x}{x^2 - 1} \tag{11.12}$$

$$f(x) = \frac{\log(2x+5)}{2x+5} \tag{11.13}$$

$$f(x) = \frac{1}{e^x + e^{-x}} \tag{11.14}$$

$$f(x) = \frac{x}{1-\sqrt{x}} \tag{11.15}$$

$$f(x) = \frac{\log(5-2x)}{5-2x} \tag{11.16}$$

$$f(x) = \frac{x+1}{\log(x+1)} \tag{11.17}$$

$$f(x) = \left(e^{-x} - e^{x+1}\right)^2 \tag{11.18}$$

$$f(x) = \frac{1 - \sqrt{x-1}}{x-1} \tag{11.19}$$

$$f(x) = \frac{x}{(x-2)^2} \tag{11.20}$$

$$f(x) = \frac{x+1}{(x-1)^2} \tag{11.21}$$

$$f(x) = \frac{1-x}{(x-3)^2} \tag{11.22}$$

$$f(x) = \frac{e^x}{(1-x)^2} \tag{11.23}$$

$$f(x) = \frac{e^{-x}}{(x-2)^2} \tag{11.24}$$

$$f(x) = \frac{-e^{-x}}{(1+x)^2} \tag{11.25}$$

$$f(x) = \frac{-e^x}{(3-x)^2} \tag{11.26}$$

$$f(x) = \frac{2\sqrt{x} - 1}{x} \tag{11.27}$$

$$f(x) = \left(e^x + e^{-x}\right)^3 \tag{11.28}$$

$$f(x) = \frac{x}{1 - \log x} \tag{11.29}$$

$$f(x) = \frac{1 - \log(x-1)}{x-1} \tag{11.30}$$

$$f(x) = \log(\sin x) \tag{11.31}$$

11.2. SOLUTIONS

11.2.1. Problem 11.1

Study the function:

$$f(x) = x^3$$

11.2.1.1. Domain of the Function

The domain of the function is $\boldsymbol{R}$:

$$f : \boldsymbol{R} \rightarrow \boldsymbol{R}$$

11.2.1.2. Intersection with the Axes

In the study of functions, the first step is to verify whether the function intersects the x, y axes:

$$x = 0 \quad \rightarrow \quad f(0) = ?$$
$$y = f(x) = 0 \quad \rightarrow \quad x = ?$$

In the present case, the function intersects the axes at the origin:

$$x = 0 \quad \rightarrow \quad y = f(0) = 0^3 = 0$$
$$y = f(x) = x^3 = 0 \quad \rightarrow \quad x = 0$$

11.2.1.3. Asymptotes

Let's verify if the function has asymptotes.

There is a vertical asymptote if (Eq. 6.41):

$$\lim_{x \to x_0} f(x) = \pm\infty, \; x_0 \in D(A)$$

The function is continuous on $\boldsymbol{R}$, so:

$$\lim_{x \to x_0} f(x) = f(x_0)$$

The function has no vertical asymptotes, since $+\infty$ and $-\infty$ do not belong to ***R***, so it cannot be $f(x_0) = \pm\infty$.

There is a horizontal asymptote if (Eq. 6.42):

$$\lim_{x \to +\infty} f(x) = l_1 \qquad l_1 \in \boldsymbol{R}$$

$$\lim_{x \to -\infty} f(x) = l_2 \qquad l_2 \in \boldsymbol{R}$$

We have:

$$\lim_{x \to +\infty} f(x) = \lim_{x \to +\infty} x^3 = +\infty$$

$$\lim_{x \to -\infty} f(x) = \lim_{x \to -\infty} x^3 = -\infty$$

The function has no horizontal asymptotes.

Since there are no horizontal asymptotes, we have to verify if there are oblique asymptotes (Eq. 6.45 – 6.48):

$$\lim_{x \to +\infty} \frac{f(x)}{x} = m_1 \qquad m_1 \in \boldsymbol{R}\text{ , } \lim_{x \to +\infty} \frac{f(x)}{x} = m_2 \qquad m_2 \in \boldsymbol{R}$$

In this case:

$$\lim_{x \to +\infty} \frac{x^3}{x} = \lim_{x \to +\infty} x^2 = +\infty$$

$$\lim_{x \to -\infty} \frac{x^3}{x} = \lim_{x \to -\infty} x^2 = +\infty$$

There is no oblique asymptote. The function has no asymptotes.

11.2.1.4. Maximum and/or Minimum Points

We verify if the function has maximum or minimum points, checking if the first derivative is equal to 0 at some point and if for these points the second derivative is > 0 or < 0.

$$f'(x_i) = 0$$

$$f''(x_i) > 0 \quad or \quad f''(x_i) < 0$$

First derivative:

$$f'(x) = \frac{d}{dx}\left(x^3\right) = 3x^2$$

The first derivative is equal to 0 at the point $x = 0$:

$$f'(x) = 0 \quad \Leftrightarrow \quad x = 0$$

Second derivative:

$$f''(x) = \frac{d}{dx}\left(f'(x)\right) = \frac{d}{dx}\left(3x^2\right) = 6x$$

The second derivative, at the point $x = 0$ where the first derivative is equal to 0, it is:

$$f''(0) = 6 \cdot 0 = 0$$

In this case there is a condition of ambiguity, since:

$f'(0) = 0$ condition for maximum or minimum point

$f''(0) = 0$ condition for inflection point

11.2.1.5. Inflection Points

In general, beyond the found ambiguity condition that will be addressed specifically, we verify whether the function has any other inflection points, checking whether the second derivative is equal to 0 at other points $x \neq 0$.

Since it holds:

$$f''(x) \neq 0, \quad \forall x \neq 0$$

there are no other points for which the second derivative is = 0.

To understand if $x_0 = 0$ is a maximum/minimum or inflection point, it is necessary to verify the curvature of the function in a neighborhood of the point $x_0 = 0$:

$$\begin{matrix} f''(x) > 0, & x > x_0 \\ f''(x) > 0 & x < x_0 \end{matrix} \quad \Rightarrow x_0 \text{ is a point of minimum}$$

$$\begin{matrix} f''(x) < 0, & x > x_0 \\ f''(x) < 0 & x < x_0 \end{matrix} \quad \Rightarrow x_0 \text{ is a point of maximum}$$

$$\begin{matrix} f''(x) > 0, & x > x_0 \\ f''(x) < 0 & x < x_0 \end{matrix} \quad \Rightarrow x_0 \text{ is an inflection point}$$

$$\begin{matrix} f''(x) < 0, & x > x_0 \\ f''(x) > 0 & x < x_0 \end{matrix} \quad \Rightarrow x_0 \text{ is an inflection point}$$

In the present case:

$$\begin{matrix} f''(x) = 6x > 0, & x > 0 \\ f''(x) = 6x < 0 & x < 0 \end{matrix} \quad \Rightarrow 0 \text{ is an inflection point}$$

It remains to calculate the value that assumes the function at the inflection point:

$$f(0) = 0^3 = 0$$

The graph of the function is shown in Figure 11.1.

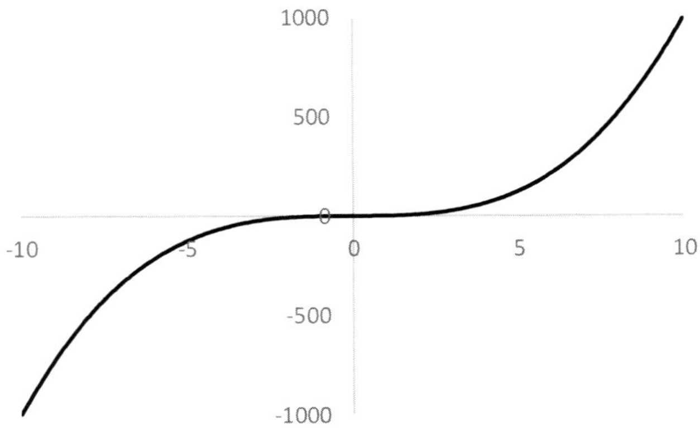

Figure 11.1. Graph of the function $f(x) = x^3$.

11.2.2. Problem 11.2

Study the function:

$$f(x) = x^4$$

11.2.2.1. Domain of the function

The domain of the function is ***R***:

$$f : \boldsymbol{R} \rightarrow \boldsymbol{R}$$

11.2.2.1. Intersection with the Axes

We verify whether the function intersects the *x*, *y* axes:

$$x = 0 \quad \rightarrow \quad f(0) = ?$$
$$y = f(x) = 0 \quad \rightarrow \quad x = ?$$

In the present case, the function intersects the axes at the origin:

$$x = 0 \quad \rightarrow \quad y = f(0) = 0^4 = 0$$
$$y = f(x) = x^4 = 0 \quad \rightarrow \quad x = 0$$

11.2.2.2. Asymptotes

Let's verify if the function has asymptotes. There is a vertical asymptote if:

$$\lim_{x \to x_0} f(x) = \pm\infty$$

The function is continuous on ***R***, so:

$$\lim_{x \to x_0} f(x) = f(x_0)$$

The function has no vertical asymptotes, since $+\infty$ and $-\infty$ do not belong to ***R***, so it cannot be $f(x_0) = \pm\infty$.

There is a horizontal asymptote if:

$$\lim_{x\to+\infty} f(x) = l_1 \qquad l_1 \in \boldsymbol{R}$$

$$\lim_{x\to-\infty} f(x) = l_2 \qquad l_2 \in \boldsymbol{R}$$

We have:

$$\lim_{x\to+\infty} f(x) = \lim_{x\to+\infty} x^4 = +\infty$$

$$\lim_{x\to-\infty} f(x) = \lim_{x\to-\infty} x^4 = +\infty$$

The function has no horizontal asymptotes.

Since there are no horizontal asymptotes, we have to verify if there are oblique asymptotes:

$$\lim_{x\to+\infty} \frac{f(x)}{x} = m_1 \qquad m_1 \in \boldsymbol{R}, \lim_{x\to+\infty} \frac{f(x)}{x} = m_2 \qquad m_2 \in \boldsymbol{R}$$

In this case:

$$\lim_{x\to+\infty} \frac{x^4}{x} = \lim_{x\to+\infty} x^3 = +\infty$$

$$\lim_{x\to-\infty} \frac{x^4}{x} = \lim_{x\to-\infty} x^3 = -\infty$$

There is no oblique asymptote. The function has no asymptotes.

11.2.2.3. Maximum and/or Minimum Points

We verify if the function has maximum or minimum points, checking if the first derivative is equal to 0 at some point and if for these points the second derivative is > 0 or < 0.

$$f'(x_i) = 0$$

$$f''(x_i) > 0 \qquad or \qquad f''(x_i) < 0$$

First derivative:

$$f'(x) = \frac{d}{dx}\left(x^4\right) = 4x^3$$

The first derivative is equal to 0 at the point $x = 0$:

$$f'(x) = 0 \quad \Leftrightarrow \quad x = 0$$

Second derivative:

$$f''(x) = \frac{d}{dx}(f'(x)) = \frac{d}{dx}\left(4x^3\right) = 12x^2$$

The second derivative, at the point $x = 0$ where the first derivative is equal to 0, it is:

$$f''(0) = 12 \cdot 0 = 0$$

In this case there is a condition of ambiguity, since:

$f'(0) = 0$ condition for maximum or minimum point

$f''(0) = 0$ condition for inflection point

11.2.2.4. Inflection Points

In general, beyond the found ambiguity condition that will be addressed specifically, we verify whether the function has any other inflection points, checking whether the second derivative is equal to 0 at other points $x \neq 0$.

Since it holds:

$$f''(x) \neq 0, \qquad \forall x \neq 0$$

there are no other points for which the second derivative is = 0.

To understand if $x_0 = 0$ is a maximum/minimum or inflection point, it is necessary to verify the curvature of the function in a neighborhood of the point $x_0 = 0$.

In the present case:

$$\begin{array}{ll} f''(x) = 12x^2 > 0, \quad x > 0 & \\ f''(x) = 12x^2 > 0 \quad x < 0 & \end{array} \Rightarrow 0 \text{ is a point of minimum}$$

The value that assumes the function at the minimum point is:

$$f(0) = 0^4 = 0$$

The graph of the function is shown in Figure 11.2.

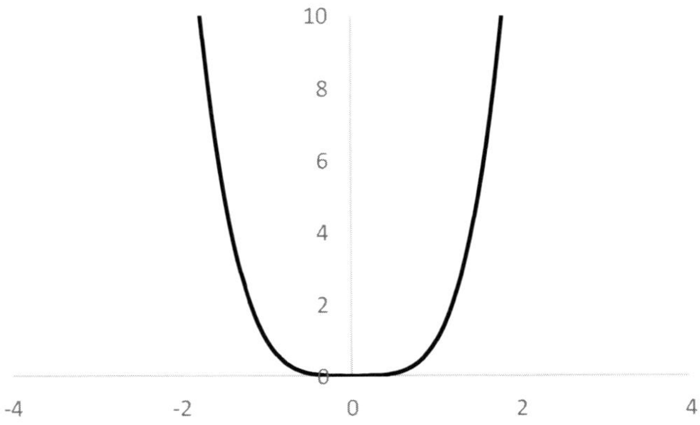

Figure 11.2. Graph of the function $f(x) = x^4$.

11.2.2.5. Comments on the Functions of the Problems 11.1 and 11.2

In a general case of function:

$$f(x) = x^n \quad ,n \in N \; n \text{ even or odd}$$

The behavior of the function at point $x_0 = 0$ is similar to what we have seen: for n odd the value 0 is an inflection point, for n even the value 0 is a minimum point.

In fact, for an odd exponent:

$$f(x) = x^{2n+1} \quad ,n \in N$$

$$f'(x) = \frac{d}{dx}\left(x^{2n+1}\right) = (2n+1)x^{2n}$$

$$f''(x) = \frac{d}{dx}\left(f'(x)\right) = \frac{d}{dx}\left((2n+1)x^{2n}\right) = 2n \cdot (2n+1)x^{2n-1}$$

$f'(0) = 0$ condition for maximum or minimum point

$f''(0) = 0$ condition for inflection point

To understand if $x_0 = 0$ is a maximum/minimum or inflection point, it is necessary to verify the curvature of the function in a neighborhood of the point $x_0 = 0$. In the present case:

$$\begin{aligned} f''(x) &= 2n\cdot(2n+1)x^{2n-1} > 0, \quad x>0 \\ f''(x) &= 2n\cdot(2n+1)x^{2n-1} < 0 \quad x<0 \end{aligned} \quad \Rightarrow \text{0 is an inflection point}$$

For an even exponent:

$$f(x) = x^{2n} \quad , \quad n \in N$$

$$f'(x) = \frac{d}{dx}\left(x^{2n}\right) = 2n \cdot x^{2n-1}$$

$$f''(x) = \frac{d}{dx}\left(f'(x)\right) = \frac{d}{dx}\left(2n \cdot x^{2n-1}\right) = 2n\cdot(2n-1)x^{2n-2}$$

$f'(0) = 0$ condition for maximum or minimum point

$f''(0) = 0$ condition for inflection point

To understand if $x_0 = 0$ is a maximum/minimum or inflection point, it is necessary to verify the curvature of the function in a neighborhood of the point $x_0 = 0$. In the present case:

$$\begin{aligned} f''(x) &= 2n\cdot(2n-1)x^{2n-2} > 0, \quad x>0 \\ f''(x) &= 2n\cdot(2n-1)x^{2n-2} > 0 \quad x<0 \end{aligned} \quad \Rightarrow \text{0 is a point of minimum}$$

11.2.3. Problem 11.3

Study the function:

$$f(x) = \frac{x^4}{4} - \frac{x^3}{3}$$

11.2.3.1. Domain of the Function

The domain of the function is ***R***:

$f:\boldsymbol{R}\rightarrow\boldsymbol{R}$

11.2.3.2. Intersection with the Axes

We verify whether the function intersects the *x*, *y* axes:

$$x=0 \quad \rightarrow \quad f(0)=?$$
$$y=f(x)=0 \quad \rightarrow \quad x=?$$

Intersection with the *y*-axis:

$$x=0 \quad \rightarrow \quad \frac{0^4}{4}-\frac{0^3}{3}=0 \quad \rightarrow \quad y=0$$

For $x = 0$, the function is equal to 0.

Intersection with the *x*-axis:

$$y=0 \quad \rightarrow \quad \frac{x^4}{4}-\frac{x^3}{3}=0 \quad \rightarrow \quad x^3\left(\frac{x}{4}-\frac{1}{3}\right)=0$$

In addition to $x = 0$, the function is = 0 for:

$$\frac{x}{4}-\frac{1}{3}=0 \quad \rightarrow \quad \frac{x}{4}=\frac{1}{3} \quad \rightarrow \quad x=\frac{4}{3}$$

The function intersects both axes at the origin (0,0) and the *x*-axis for $x=\frac{4}{3}$.

11.2.3.3. Asymptotes

There is a vertical asymptote if:

$$\lim_{x\rightarrow x_0} f(x)=\pm\infty$$

The function is continuous on $\boldsymbol{R}$, so:

$$\lim_{x\rightarrow x_0} f(x)=f(x_0)$$

The function has no vertical asymptotes, since $+\infty$ and $-\infty$ do not belong to $\boldsymbol{R}$, so it cannot be $f(x_0) = \pm\infty$.

There is a horizontal asymptote if:

$$\lim_{x\to+\infty} f(x) = l_1 \qquad l_1 \in \boldsymbol{R}$$

$$\lim_{x\to-\infty} f(x) = l_2 \qquad l_2 \in \boldsymbol{R}$$

We have:

$$\lim_{x\to+\infty} f(x) = \lim_{x\to+\infty}\left(\frac{x^4}{4} - \frac{x^3}{3}\right) = \lim_{x\to+\infty} x^3\left(\frac{x}{4} - \frac{1}{3}\right) = (+\infty)(+\infty) = +\infty$$

$$\lim_{x\to-\infty} f(x) = \lim_{x\to-\infty} x^3\left(\frac{x}{4} - \frac{1}{3}\right) = (-\infty)(-\infty) = +\infty$$

The function has no horizontal asymptotes.

Since there are no horizontal asymptotes, we have to verify if there are oblique asymptotes:

$$\lim_{x\to+\infty} \frac{f(x)}{x} = m_1 \qquad m_1 \in \boldsymbol{R}\text{, } \lim_{x\to+\infty} \frac{f(x)}{x} = m_2 \qquad m_2 \in \boldsymbol{R}$$

In this case:

$$\lim_{x\to+\infty} \frac{x^3\left(\frac{x}{4} - \frac{1}{3}\right)}{x} = \lim_{x\to+\infty} x^2\left(\frac{x}{4} - \frac{1}{3}\right) = (+\infty)(+\infty) = +\infty$$

$$\lim_{x\to-\infty} \frac{x^3\left(\frac{x}{4} - \frac{1}{3}\right)}{x} = \lim_{x\to-\infty} x^2\left(\frac{x}{4} - \frac{1}{3}\right) = (+\infty)(-\infty) = -\infty$$

There is no oblique asymptote. The function has no asymptotes.

11.2.3.4. Maximum and/or Minimum Points

We verify if the function has maximum or minimum points, checking if the first derivative is equal to 0 at some point and if for these points the second derivative is > 0 or < 0.

$$f'(x_i) = 0$$

$$f''(x_i) > 0 \quad or \quad f''(x_i) < 0$$

First derivative:

$$f'(x) = \frac{d}{dx}\left(\frac{x^4}{4} - \frac{x^3}{3}\right) = \frac{4x^3}{4} - \frac{3x^2}{3} = x^3 - x^2 = x^2(x-1)$$

The first derivative is equal to 0 at the points:

$$f'(x) = 0 \quad \Leftrightarrow \quad \begin{matrix} x = 0 \\ x = 1 \end{matrix}$$

Second derivative:

$$f''(x) = \frac{d}{dx}\left(x^3 - x^2\right) = 3x^2 - 2x = x(3x-2)$$

The second derivative, at the points where the first derivative is equal to 0, it is:

$$f''(0) = 0$$

$f''(1) = 3 - 2 = 1 > 0$ 1 is a point of minimum

$x = 0$ can be an inflection point, to be verified.

11.2.3.5. Inflection Points

We verify where the second derivative is equal to 0:

$$f''(x) = x(3x-2) = 0$$

The second derivative is equal to 0 in the points:

$x = 0$ eventual inflection point to be checked

$$3x - 2 = 0 \quad \rightarrow \quad x = \frac{2}{3} \text{ inflection point}$$

It holds:

$$\begin{aligned} f''(x) &= x(3x-2) > 0, \quad x < 0 \\ f''(x) &= x(3x-2) < 0, \quad 0 < x < \frac{2}{3} \end{aligned} \quad \Rightarrow 0 \text{ is an inflection point.}$$

The values of the function at the minimum and inflection points are:

- point of minimum:

$$f(1) = \frac{1}{4} - \frac{1}{3} = -\frac{1}{12}$$

- inflection point:

$$f(0) = \frac{0^4}{4} - \frac{0^3}{3} = 0$$

- inflection point:

$$f\left(\frac{2}{3}\right) = \frac{1}{4}\left(\frac{2}{3}\right)^4 - \frac{1}{3}\left(\frac{2}{3}\right)^3 = \frac{16}{4 \cdot 81} - \frac{8}{3 \cdot 27} = \frac{4}{81} - \frac{8}{81} = -\frac{4}{81}$$

The graph of the function is shown in Figure 11.3.

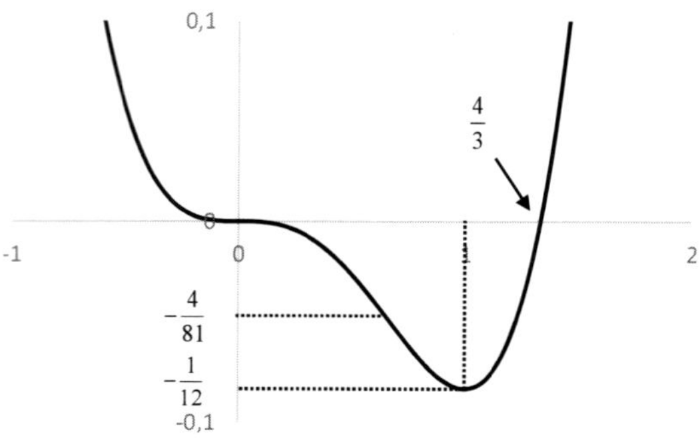

Figure 11.3. Graph of the function $f(x) = \frac{x^4}{4} - \frac{x^3}{3}$.

11.2.4. Problem 11.4

Study the function:

$$f(x) = \frac{x-1}{x^2}$$

11.2.4.1. Domain of the Function

The domain of definition of the function is $\boldsymbol{R}$ (both in the numerator and in the denominator) minus the values in which the denominator is equal to zero:

$$D = \boldsymbol{R} - \{0\}$$

The function is therefore defined by:

$$f : \boldsymbol{R} - \{0\} \to \boldsymbol{R}$$

11.2.4.2. Intersection with the Axes

We verify whether the function intersects the x, y axes:

$$x = 0 \quad \to \quad f(0) = ?$$
$$y = f(x) = 0 \quad \to \quad x = ?$$

In this case, f is not defined for $x = 0$, so the function does not intersect the y axis.
Intersection with the x-axis:

$$f(x) = \frac{x-1}{x^2} = 0$$

We have:

$$y = f(x) = 0 \quad \Leftrightarrow \quad x - 1 = 0 \quad \Rightarrow \quad x = 1$$

The function intersects the x-axis for $x = 1$.

11.2.4.3. Asymptotes

There is a vertical asymptote if:

$$\lim_{x\to 0+}\frac{x-1}{x^2}=\pm\infty$$

$$\lim_{x\to 0-}\frac{x-1}{x^2}=\pm\infty$$

In particular, the denominator always tends to + ∞:

$$\lim_{x\to 0}\frac{1}{x^2}=\lim_{x\to 0+}\frac{1}{x^2}=\lim_{x\to 0-}\frac{1}{x^2}=+\infty$$

The sign is determined by the numerator:

$$\lim_{x\to 0+}(x-1)=\lim_{x\to 0-}(x-1)=-1$$

for which:

$$\lim_{x\to 0+}\frac{x-1}{x^2}=-\infty$$

$$\lim_{x\to 0-}\frac{x-1}{x^2}=-\infty$$

There is a vertical asymptote at $x = 0$ with two infinite branches to – ∞.
There is a horizontal asymptote if:

$$\lim_{x\to -\infty}\frac{x-1}{x^2}=l_1,\ l_1\in \boldsymbol{R}$$

$$\lim_{x\to +\infty}\frac{x-1}{x^2}=l_2,\ l_2\in \boldsymbol{R}$$

We have:

$$\lim_{x\to -\infty}\frac{x-1}{x^2}=\frac{-\infty}{+\infty}$$

It is an indeterminate form, we can apply De L'Hôpital's rule (Eq. 5.35 and 5.37).

$$\lim_{x\to-\infty}\frac{x-1}{x^2}=\lim_{x\to-\infty}\frac{\frac{d}{dx}(x-1)}{\frac{d}{dx}(x^2)}=\lim_{x\to-\infty}\frac{1}{2x}=0$$

$$\lim_{x\to+\infty}\frac{x-1}{x^2}=\lim_{x\to+\infty}\frac{\frac{d}{dx}(x-1)}{\frac{d}{dx}(x^2)}=\lim_{x\to+\infty}\frac{1}{2x}=0$$

There is a horizontal asymptote at $y = 0$ and the function has two infinite branches, one for each half-plane ($x > 0$, $x < 0$) with respect to the y-axis. For $x < 0$ the function tends to 0 by negative values, for $x > 0$ the function tends to 0 by positive values.

11.2.4.4. Maximum and/or Minimum Points

We verify if the function has maximum or minimum points, checking if the first derivative is equal to 0 at some point and if for these points the second derivative is > 0 or < 0.

$$f'(x_i)=0$$
$$f''(x_i)>0 \quad or \quad f''(x_i)<0$$

First derivative:

$$f'(x)=\frac{d}{dx}\left(\frac{x-1}{x^2}\right)=\frac{1\cdot x^2-(x-1)\cdot 2x}{x^4}=\frac{x^2-2x^2+2x}{x^4}=$$

$$=\frac{x\cdot(2-x)}{x^4}=\frac{2-x}{x^3}$$

The first derivative is equal to 0 when the numerator is equal to 0:

$$f'(x)=0 \quad \Leftrightarrow \quad 2-x=0 \quad \to \quad x=2$$

Second derivative:

$$f''(x) = \frac{d}{dx}\left(f'(x)\right) = \frac{d}{dx}\left(\frac{2-x}{x^3}\right) = \frac{-1\cdot x^3 - (2-x)\cdot 3x^2}{x^6} =$$

$$= \frac{-x^3 - 6x^2 + 3x^3}{x^6} = \frac{2x^3 - 6x^2}{x^6} = \frac{2(x-3)}{x^4}$$

The second derivative, at the point $x = 2$ where the first derivative is equal to 0, it is:

$$f''(2) = \frac{2(2-3)}{2^4} < 0$$

$x = 2$ is a point of maximum

11.2.4.5. Inflection Points

We verify if the second derivative is equal to 0:

$$f''(x) = \frac{2(x-3)}{x^4} = 0$$

The second derivative is equal to 0 when the numerator is equal to 0:

$$f''(x) = 0 \quad \Leftrightarrow \quad x - 3 = 0 \quad \rightarrow \quad x = 3$$

$x = 3$ is an inflection point.

The values of the function at the maximum and inflection points are:

$f(2) = \frac{2-1}{2^2} = \frac{1}{4}$ (2 is a point of maximum)

$f(3) = \frac{3-1}{3^2} = \frac{2}{9}$ (3 is an inflection point)

The graph of the function is shown in Figure 11.4.

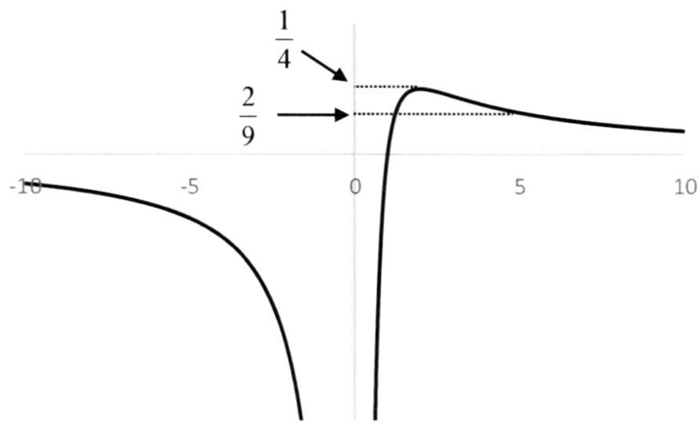

Figure 11.4. Graph of the function $f(x) = \dfrac{x-1}{x^2}$.

11.2.5. Problem 11.5

Study the function:

$$f(x) = x^2 e^x$$

11.2.5.1. Domain of the Function

The domain of the function is $\boldsymbol{R}$:

$$f : \boldsymbol{R} \to \boldsymbol{R}$$

11.2.5.2. Intersection with the Axes

We verify whether the function intersects the *x*, *y* axes:

$$x = 0 \quad \to \quad f(0) = ?$$

$$y = f(x) = 0 \quad \to \quad x = ?$$

In the present case:

$$x = 0 \quad \to \quad 0^2 e^0 = 0 \quad \to \quad y = 0$$

$$y = x^2 e^x = 0 \Leftrightarrow x = 0 \to x = 0$$

The function intersects the axes at the origin (0,0).

11.2.5.3. Asymptotes

There is a vertical asymptote if:

$$\lim_{x \to x_0+} x^2 e^x = \pm\infty$$

$$\lim_{x \to x_0-} x^2 e^x = \pm\infty$$

There is no point where at least one of the above conditions occurs. The function is continuous, so:

$$\lim_{x \to x_0} x^2 e^x = x_0^2 e^{x_0}$$

if $x_0 \in \boldsymbol{R}$, the function value cannot be $\pm\infty$.

There is a horizontal asymptote if:

$$\lim_{x \to -\infty} x^2 e^x = l_1 \quad l_1 \in \boldsymbol{R}$$

$$\lim_{x \to +\infty} x^2 e^x = l_2 \quad l_2 \in \boldsymbol{R}$$

We have:

$$\lim_{x \to -\infty} x^2 e^x = +\infty \cdot 0$$

It is an indeterminate form, the limit can be written as:

$$\lim_{x \to -\infty} x^2 e^x = \lim_{x \to -\infty} \frac{x^2}{e^{-x}} = \frac{+\infty}{+\infty}$$

It is still an indeterminate form, we can apply De L'Hôpital's rule:

$$\lim_{x \to -\infty} x^2 e^x = \lim_{x \to -\infty} \frac{x^2}{e^{-x}} = \lim_{x \to -\infty} \frac{\frac{d}{dx}\left(x^2\right)}{\frac{d}{dx}\left(e^{-x}\right)} = \lim_{x \to -\infty} \frac{2x}{-e^{-x}} = \frac{+\infty}{-\infty}$$

After the first derivation the limit still has an indeterminate form, so we can continue to apply the De L'Hôpital's rule.

$$\lim_{x \to -\infty} \frac{2x}{-e^{-x}} = \lim_{x \to -\infty} \frac{\frac{d}{dx}(2x)}{\frac{d}{dx}\left(-e^{-x}\right)} = \lim_{x \to -\infty} \frac{2}{e^{-x}} = 0$$

For $x \to -\infty$ there is a horizontal asymptote at $y = 0$:

$$\lim_{x \to -\infty} x^2 e^x = 0$$

Let's check if there is a horizontal asymptote also for $x \to +\infty$:

$$\lim_{x \to +\infty} x^2 e^x = +\infty$$

There is no horizontal asymptote for $x \to +\infty$, so it is necessary to check whether there is an oblique asymptote (Eq. 6.45):

$$\lim_{x \to +\infty} \frac{f(x)}{x} = m$$

In this case:

$$\lim_{x \to +\infty} \frac{x^2 e^x}{x} = \lim_{x \to +\infty} x e^x = +\infty$$

There is no oblique asymptote, there is only one horizontal asymptote at $y = 0$ and the function has an infinite branch in the half-plane $x < 0$.

11.2.5.4. Maximum and/or Minimum Points

We verify if the function has maximum or minimum points, checking if the first derivative is equal to 0 at some point and if for these points the second derivative is > 0 or < 0.

$$f'(x_i) = 0$$

$$f''(x_i) > 0 \quad or \quad f''(x_i) < 0$$

First derivative:

$$f'(x) = \frac{d}{dx}\left(x^2 e^x\right) = 2x \cdot e^x + x^2 \cdot e^x = e^x\left(x^2 + 2x\right) \equiv xe^x(x+2)$$

$$f'(x) = 0 \quad \Leftrightarrow \quad \begin{matrix} x = 0 \\ x = -2 \end{matrix}$$

The first derivative is equal to 0 at the points $x = 0$ and $x = -2$.

Second derivative:

$$f''(x) = \frac{d}{dx}\left(f'(x)\right) = \frac{d}{dx}\left(e^x\left(x^2 + 2x\right)\right) =$$
$$= e^x\left(x^2 + 2x\right) + e^x(2x+2) = e^x\left(x^2 + 4x + 2\right)$$

The second derivative, at the points where the first derivative is equal to 0, it is:

$$f''(x) = e^x\left(x^2 + 4x + 2\right)$$
$$f''(0) = 2e^0 = 2 > 0 \ \ 0 \text{ is a point of minimum}$$
$$f''(-2) = e^{-2}(4 - 8 + 2) = -2e^{-2} < 0 \ \ -2 \text{ is a point of maximum}$$

11.2.5.5. Inflection Points

We verify whether the function has inflection points, checking whether the second derivative is equal to 0.

$$f''(x) = e^x\left(x^2 + 4x + 2\right) = 0$$

The second derivative is equal to 0 when:

$$x^2 + 4x + 2 = 0$$

A second degree equation:

$$ax^2 + bx + c = 0$$

it has the solutions:

$$\frac{-b \pm \sqrt{b^2 - 4ac}}{2a}$$

In the present case:

$$a = 1 \qquad b = 4 \qquad c = 2$$

The roots of the equation are:

$$\frac{-4 \pm \sqrt{16-8}}{2} = \frac{-4 \pm \sqrt{8}}{2} = \frac{-4 \pm 2\sqrt{2}}{2} = \frac{2\left(-2 \pm \sqrt{2}\right)}{2} = -2 \pm \sqrt{2}$$

So the function has two inflection points at the values:

$$x = -2 + \sqrt{2}$$
$$x = -2 - \sqrt{2}$$

The values of the function at the maximum, minimum and inflection points are:
$x = 0$ is a point of minimum:

$$f(0) = 0^2 e^0 = 0$$

$x = -2$ is a point of maximum:

$$f(-2) = (-2)^2 e^{-2} = 4e^{-2}$$

$x = -2 + \sqrt{2}$ is an inflection point:

$$f(-2+\sqrt{2}) = \left(-2+\sqrt{2}\right)^2 e^{-2+\sqrt{2}} = \left(6 - 4\sqrt{2}\right)e^{-2+\sqrt{2}} = \left(6 - 4\sqrt{2}\right)e^{-\left(2-\sqrt{2}\right)}$$

$x = -2 - \sqrt{2}$ is an inflection point:

$$f(-2-\sqrt{2}) = \left(-2-\sqrt{2}\right)^2 e^{-2-\sqrt{2}} = \left(6 + 4\sqrt{2}\right)e^{-2-\sqrt{2}} = \left(6 + 4\sqrt{2}\right)e^{-\left(2+\sqrt{2}\right)}$$

The graph of the function is shown in Figure 11.5.

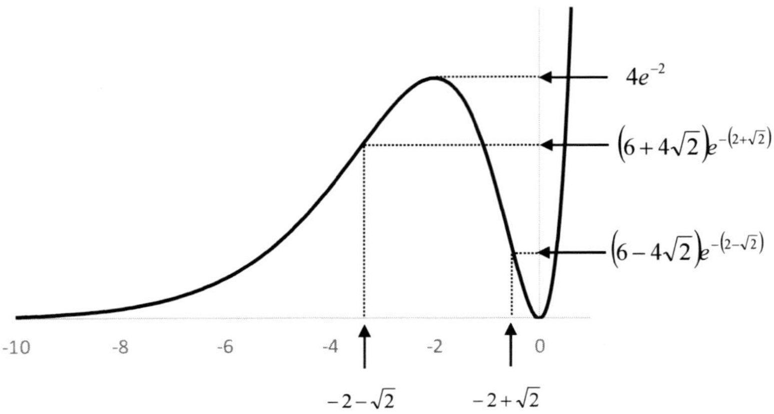

Figure 11.5. Graph of the function $f(x) = x^2 e^x$.

11.2.5.6. Appendix

In the values of the function they appear both square roots and exponentials, so it would need adequate tools (e.g., a calculator) to correctly calculate the values. Otherwise, it is necessary at least to establish which of the two values is greater than the other to draw the graph correctly.

That is, it is necessary to establish whether:

$$f(-2+\sqrt{2}) = \left(6-4\sqrt{2}\right)e^{-2+\sqrt{2}} \quad > \quad f(-2-\sqrt{2}) = \left(6+4\sqrt{2}\right)e^{-2-\sqrt{2}}$$

or:

$$f(-2+\sqrt{2}) = \left(6-4\sqrt{2}\right)e^{-2+\sqrt{2}} \quad < \quad f(-2-\sqrt{2}) = \left(6+4\sqrt{2}\right)e^{-2-\sqrt{2}}$$

Since it is not necessary to calculate values but to verify an inequality, it can be considered the ratio:

$$\frac{\left(6+4\sqrt{2}\right)\cdot e^{-2-\sqrt{2}}}{\left(6-4\sqrt{2}\right)\cdot e^{-2+\sqrt{2}}}$$

The ratio can be < 1 or > 1. Assuming it is > 1, it is sufficient to check if there is a real number "*a*" such that:

$$\frac{\left(6+4\sqrt{2}\right)\cdot e^{-2-\sqrt{2}}}{\left(6-4\sqrt{2}\right)\cdot e^{-2+\sqrt{2}}} > a > 1$$

It holds:

$$\frac{\left(6+4\sqrt{2}\right)\cdot e^{-2-\sqrt{2}}}{\left(6-4\sqrt{2}\right)\cdot e^{-2+\sqrt{2}}}=\frac{\left(6+4\sqrt{2}\right)\cdot e^{-2}\cdot e^{-\sqrt{2}}}{\left(6-4\sqrt{2}\right)\cdot e^{-2}\cdot e^{+\sqrt{2}}}=\frac{\left(6+4\sqrt{2}\right)\cdot e^{-\sqrt{2}}}{\left(6-4\sqrt{2}\right)\cdot e^{+\sqrt{2}}}=$$

$$=\frac{\left(6+4\sqrt{2}\right)}{\left(6-4\sqrt{2}\right)\cdot e^{+\sqrt{2}}\cdot e^{+\sqrt{2}}}=\frac{\left(6+4\sqrt{2}\right)}{\left(6-4\sqrt{2}\right)}\cdot\frac{1}{e^{+\sqrt{2}+\sqrt{2}}}=\frac{\left(6+4\sqrt{2}\right)}{\left(6-4\sqrt{2}\right)}\cdot\frac{1}{e^{2\sqrt{2}}}$$

Working on the previous ratio, we obtain:

$$\frac{\left(6+4\sqrt{2}\right)}{\left(6-4\sqrt{2}\right)}\cdot\frac{1}{e^{2\sqrt{2}}}=\frac{\left(6+4\sqrt{2}\right)\cdot\left(6+4\sqrt{2}\right)}{\left(6-4\sqrt{2}\right)\cdot\left(6+4\sqrt{2}\right)}\cdot\frac{1}{e^{2\sqrt{2}}}=\frac{\left(6+4\sqrt{2}\right)^2}{6^2-\left(4\sqrt{2}\right)^2}\cdot\frac{1}{e^{2\sqrt{2}}}=$$

$$=\frac{36+32+48\sqrt{2}}{36-32}\cdot\frac{1}{e^{2\sqrt{2}}}=\frac{68+48\sqrt{2}}{4}\cdot\frac{1}{e^{2\sqrt{2}}}=\frac{4\left(17+12\sqrt{2}\right)}{4}\cdot\frac{1}{e^{2\sqrt{2}}}=$$

$$=\left(17+12\sqrt{2}\right)\cdot\frac{1}{e^{2\sqrt{2}}}$$

Having assumed that the ratio is > 1, we have:

$$\left(17+12\sqrt{2}\right)\cdot\frac{1}{e^{2\sqrt{2}}}>1 \quad\Rightarrow\quad \left(17+12\sqrt{2}\right)>e^{2\sqrt{2}}$$

By using the logarithm function:

$$\log\left(17+12\sqrt{2}\right)>\log\left(e^{2\sqrt{2}}\right) \quad\Rightarrow\quad \log\left(17+12\sqrt{2}\right)>2\sqrt{2}$$

Considering that:

$$\left(17+12\sqrt{2}\right)>17+12=29$$

we obtain:

$$\log\left(17+12\sqrt{2}\right)>\log 29>\log 27=\log 3^3=3\log 3$$

It holds:

$$3\log 3 > 3\log e = 3 \quad , \quad 1.5 > \sqrt{2}$$

for which:

$$\log\left(17+12\sqrt{2}\right) > 3\log 3 > 3 > 2\sqrt{2}$$

Thus:

$$f\left(-2-\sqrt{2}\right) > f\left(-2+\sqrt{2}\right)$$

As drawn on the graph of Figure 11.5.

11.2.6. Problem 11.6

Study the function:

$$f(x) = x^2 \log x$$

11.2.6.1. Domain of the Function

The domain of the function is $\boldsymbol{R}^+$:

$$f : \boldsymbol{R}^+ \to \boldsymbol{R}$$

11.2.6.2. Intersection with the Axes

We verify whether the function intersects the x, y axes:

$$x = 0 \quad \to \quad f(0) = ?$$
$$y = f(x) = 0 \quad \to \quad x = ?$$

Given the domain, the function does not intersect the y-axis:

$$f : \boldsymbol{R}^+ \to \boldsymbol{R} \quad \Rightarrow \quad x \in \left]0, +\infty\right[$$

Intersection with the x-axis:

$$y = x^2 \log x = 0$$

We have:

$$\log 1 = 0 \quad \rightarrow \quad x = 1$$

11.2.6.3. Asymptotes

We have seen that the function does not intersect the *y*-axis, so we calculate the limit:

$$\lim_{x\to 0+} x^2 \log x = \lim_{x\to 0+} \frac{\log x}{x^{-2}} = \frac{-\infty}{+\infty}$$

It is an indeterminate form, we can apply De L'Hôpital's rule:

$$\lim_{x\to 0+} \frac{\log x}{x^{-2}} = \lim_{x\to 0+} \frac{\frac{d}{dx}(\log x)}{\frac{d}{dx}\left(x^{-2}\right)} = \lim_{x\to 0+} \frac{\frac{1}{x}}{-2x^{-3}} = \lim_{x\to 0+}\left(-\frac{1}{2}\cdot\frac{x^{-1}}{x^{-3}}\right) = 0$$

The function tends to 0 for $x \rightarrow 0+$, so there is no vertical asymptote.
Let's check if there is a horizontal asymptote. We have:

$$\lim_{x\to +\infty} x^2 \log x = +\infty$$

The function tends to $+\infty$ for $x \rightarrow +\infty$. There is no horizontal asymptote, so it is necessary to check whether there is an oblique asymptote:

$$\lim_{x\to +\infty} \frac{f(x)}{x} = m$$

In this case:

$$\lim_{x\to +\infty} \frac{x^2 \log x}{x} = \lim_{x\to +\infty} x \log x = +\infty$$

There is no oblique asymptote.

11.2.6.4. Maximum and/or Minimum Points

We verify if the function has maximum or minimum points, checking if the first derivative is equal to 0 at some point and if for these points the second derivative is > 0 or < 0.

$$f'(x_i) = 0$$
$$f''(x_i) > 0 \quad or \quad f''(x_i) < 0$$

First derivative:

$$f'(x) = \frac{d}{dx}\left(x^2 \log x\right) = 2x \cdot \log x + x^2 \cdot \frac{1}{x} = 2x\log x + x = x(2\log x + 1)$$

$$f'(x) = 0 \quad \Rightarrow x(2\log x + 1) = 0$$

A solution cannot be $x = 0$, since $x \in \left]0,+\infty\right[$. The only solution is:

$$2\log x + 1 = 0 \quad \rightarrow \quad \log x = -\frac{1}{2}$$

by applying the inverse function:

$$x = e^{\log x} = e^{-\frac{1}{2}}$$

Second derivative:

$$f''(x) = \frac{d}{dx}\left[x(2\log x + 1)\right] = (2\log x + 1) + \left(x \cdot \frac{2}{x}\right) = 2\log x + 3$$

The second derivative, at the point where the first derivative is equal to 0, it is:

$$f''\left(e^{-\frac{1}{2}}\right) = 2\log\left(e^{-\frac{1}{2}}\right) + 3 = 2 \cdot \left(-\frac{1}{2}\right) + 3 = 2 > 0$$

$x = e^{-\frac{1}{2}}$ is a point of minimum.

The value of the function at the minimum point is:

$$f\left(e^{-\frac{1}{2}}\right)=\left(e^{-\frac{1}{2}}\right)^2\log\left(e^{-\frac{1}{2}}\right)=-\frac{1}{2}e^{-1}$$

11.2.6.5. Inflection Points

We verify whether the function has inflection points, checking whether the second derivative is equal to 0.

$$f''(x)=0$$

$$2\log x+3=0 \quad \rightarrow \quad \log x=-\frac{3}{2}$$

The second derivative is equal to 0 at the point:

$$x=e^{\log x}=e^{-\frac{3}{2}}$$

The value of the function at the inflection point is:

$$f\left(e^{-\frac{3}{2}}\right)=\left(e^{-\frac{3}{2}}\right)^2\log\left(e^{-\frac{3}{2}}\right)=-\frac{3}{2}e^{-3}$$

The graph of the function is shown in Figure 11.6.

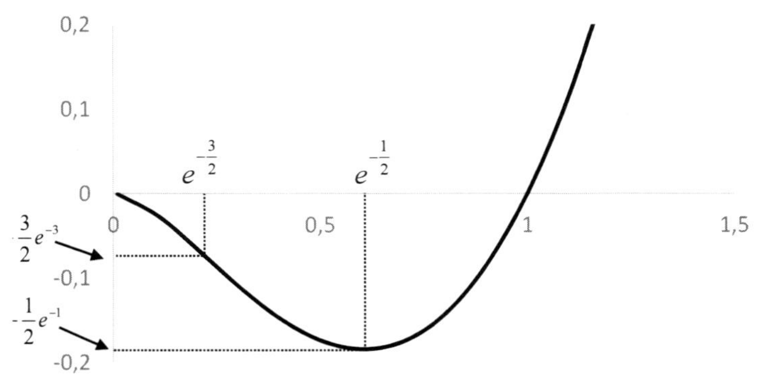

Figure 11.6. Graph of the function $f(x)=x^2\log x$.

11.2.7. Problem 11.7

Study the function:

$$f(x) = e^{-x^2}$$

11.2.7.1. Domain of the Function

The domain of the function is $\boldsymbol{R}$:

$$f : \boldsymbol{R} \to \boldsymbol{R}$$

11.2.7.2. Intersection with the Axes

We verify whether the function intersects the x, y axes:

$$x = 0 \quad \to \quad f(0) = ?$$

$$y = f(x) = 0 \quad \to \quad x = ?$$

In the present case, for $x = 0$:

$$y = f(0) = e^{-0^2} = e^0 = 1$$

The function intersects the y-axis at the point $y = 1$.
Instead, there are no x points for which it holds:

$$y = f(x) = e^{-x^2} = 0$$

So the function does not intersect the x-axis.

11.2.7.3. Asymptotes

There is a vertical asymptote if:

$$\lim_{x \to x_0+} e^{-x^2} = \pm\infty$$

$$\lim_{x \to x_0-} e^{-x^2} = \pm\infty$$

The function is continuous on $\boldsymbol{R}$, so:

$$\lim_{x \to x_0} e^{-x^2} = f(x_0) = e^{-x_0^2} \quad (x_0 \in \boldsymbol{R},\ e^{-x_0^2} \neq \pm\infty)$$

The function has no vertical asymptotes, since + ∞ and –∞ do not belong to ***R***, so it cannot be $f(x_0) = \pm\infty$.

There is a horizontal asymptote if:

$$\lim_{x \to -\infty} e^{-x^2} = l_1,\ l_1 \in \boldsymbol{R}$$

$$\lim_{x \to +\infty} e^{-x^2} = l_2,\ l_2 \in \boldsymbol{R}$$

For the function it holds:

$$\lim_{x \to -\infty} e^{-x^2} = 0$$

$$\lim_{x \to +\infty} e^{-x^2} = 0$$

There is a horizontal asymptote with two infinite branches for $y = 0$.

11.2.7.4. Maximum and/or Minimum Points

We verify if the function has maximum or minimum points, checking if the first derivative is equal to 0 at some point and if for these points the second derivative is > 0 or < 0.

$$f'(x_i) = 0$$

$$f''(x_i) > 0 \quad or \quad f''(x_i) < 0$$

First derivative:

$$f'(x) = \frac{d}{dx}\left(e^{-x^2}\right) = e^{-x^2}(-2x)$$

The first derivative is equal to 0 at the point $x = 0$.

Second derivative:

$$f''(x) = \frac{d}{dx}\left(-2xe^{-x^2}\right) = -2e^{-x^2} + (-2x)e^{-x^2}(-2x) =$$
$$= -2e^{-x^2} + 4x^2e^{-x^2} = 2e^{-x^2}\left(2x^2 - 1\right)$$

$$f''(0) = 2e^0(0-1) = -2 < 0$$

$x = 0$ is a point of maximum.

The value of the function at the maximum point is:

$$f(0) = e^{-0^2} = e^0 = 1$$

11.2.7.5. Inflection Points

We verify whether the function has inflection points, checking whether the second derivative is equal to 0.

$$f''(x) = 2e^{-x^2}\left(2x^2 - 1\right) = 0$$
$$f''(x) = 0 \quad \Leftrightarrow \quad \left(2x^2 - 1\right) = 0 \quad \rightarrow \quad x^2 = \frac{1}{2}$$

The function has two inflection points at the values:

$$x = \pm\sqrt{\frac{1}{2}}$$

The values of the function at the inflection points are:

$$f\left(-\sqrt{\frac{1}{2}}\right) = e^{-\left(-\sqrt{\frac{1}{2}}\right)^2} = e^{-\frac{1}{2}} \qquad f\left(\sqrt{\frac{1}{2}}\right) = e^{-\left(\sqrt{\frac{1}{2}}\right)^2} = e^{-\frac{1}{2}}$$

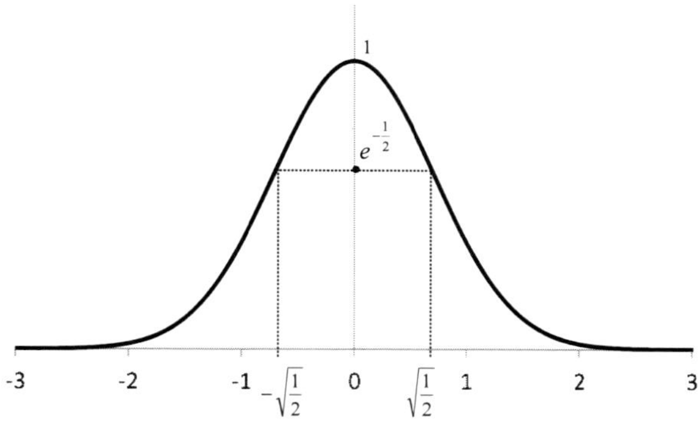

Figure 11.7. Graph of the function $f(x)=e^{-x^2}$.

The graph of the function is shown in Figure 11.7.

11.2.8. Problem 11.8

Study the function:

$$f(x)=x-\sqrt{x^2-1}$$

11.2.8.1. Domain of the Function

The domain of the function, given the presence of a square root, is determined by the values for which:

$$x^2-1\geq 0 \quad \rightarrow \quad x^2\geq 1 \quad \rightarrow \quad |x|\geq 1$$

The function is therefore defined by:

$$f:]-\infty,-1]\cup[1,+\infty[\rightarrow \boldsymbol{R}$$

11.2.8.2. Intersection with the Axes

We verify whether the function intersects the x, y axes:

$$x=0 \quad \rightarrow \quad f(0)=?$$
$$y=f(x)=0 \quad \rightarrow \quad x=?$$

The function is not defined for $x = 0$, so it does not intersect the y-axis.

For the intersection with the x-axis, it should be $y = 0$:

$$x-\sqrt{x^2-1}=0 \quad \rightarrow \quad x=\sqrt{x^2-1} \quad \rightarrow \quad x^2=x^2-1$$

The equation does not admit a solution, so the function does not intersect the x-axis.

11.2.8.3. Asymptotes

There is a vertical asymptote if:

$$\lim_{x\to x_0+}\left(x-\sqrt{x^2-1}\right)=\pm\infty$$
$$\lim_{x\to x_0-}\left(x-\sqrt{x^2-1}\right)=\pm\infty$$

Given the function definition domain, we can check the limits for $x_0 = 1$:

$$\lim_{x\to 1+}\left(x-\sqrt{x^2-1}\right)=1$$
$$\lim_{x\to -1-}\left(x-\sqrt{x^2-1}\right)=-1$$

The result is compatible with the function domain, which includes two intervals closed at the values 1 and –1.

In general, the function is continuous on the definition domain, so for any value:

$$x_0 \in \left]-\infty,-1\right]\cup\left[1,+\infty\right[$$

It holds:

$$\lim_{x\to x_0}\left(x-\sqrt{x^2-1}\right)=f(x_0)=x_0-\sqrt{x_0^2-1}$$

The function has no vertical asymptotes.

There is a horizontal asymptote if:

$$\lim_{x\to -\infty}\left(x-\sqrt{x^2-1}\right)=l_1,\ l_1\in \mathbf{R}$$
$$\lim_{x\to +\infty}\left(x-\sqrt{x^2-1}\right)=l_2,\ l_2\in \mathbf{R}$$

For the function in question it holds:

$$\lim_{x\to+\infty}\left(x-\sqrt{x^2-1}\right)=+\infty-\infty$$

It is an indeterminate form, which can be solved in the following way:

$$\lim_{x\to+\infty}\left(x-\sqrt{x^2-1}\right)=\lim_{x\to+\infty}\frac{\left(x-\sqrt{x^2-1}\right)\cdot\left(x+\sqrt{x^2-1}\right)}{\left(x+\sqrt{x^2-1}\right)}=\lim_{x\to+\infty}\frac{x^2-\left(x^2-1\right)}{x+\sqrt{x^2-1}}=$$

$$=\lim_{x\to+\infty}\frac{1}{x+\sqrt{x^2-1}}=\frac{1}{+\infty}=0$$

For $x\to+\infty$ there is a horizontal asymptote at $y=0$.
For $x\to-\infty$:

$$\lim_{x\to-\infty}\left(x-\sqrt{x^2-1}\right)=-\infty$$

For this branch there is no horizontal asymptote, so it is necessary to check if there is an oblique asymptote (Eq. 6.44, 6.46):

$$\lim_{x\to-\infty}\frac{f(x)}{x}=m;\ \lim_{x\to-\infty}\left[f(x)-mx\right]=q;\ m,q\in\boldsymbol{R}$$

$$\lim_{x\to-\infty}\frac{x-\sqrt{x^2-1}}{x}=\lim_{x\to-\infty}\left(\frac{x}{x}-\frac{\sqrt{x^2-1}}{x}\right)=\lim_{x\to-\infty}\left(1+\frac{\sqrt{x^2-1}}{-x}\right)=$$

$$=\lim_{x\to-\infty}\left(1+\sqrt{\frac{x^2-1}{x^2}}\right)=\lim_{x\to-\infty}\left(1+\sqrt{1-\frac{1}{x^2}}\right)=2$$

There is an oblique asymptote and the slope of the straight line is:

$$m=2$$

q is the y-intercept and it is given by:

$$q = \lim_{x\to-\infty}[f(x) - mx] = \lim_{x\to-\infty}\left[\left(x - \sqrt{x^2-1}\right) - 2x\right] = \lim_{x\to-\infty}\left(-x - \sqrt{x^2-1}\right) =$$

$$= \lim_{x\to-\infty} -\left(x + \sqrt{x^2-1}\right) = \lim_{x\to-\infty} - \frac{\left(x+\sqrt{x^2-1}\right)\cdot\left(x-\sqrt{x^2-1}\right)}{x-\sqrt{x^2-1}} =$$

$$= \lim_{x\to-\infty} - \frac{x^2-\left(x^2-1\right)}{x-\sqrt{x^2-1}} = \lim_{x\to-\infty} - \frac{1}{x-\sqrt{x^2-1}} = -\frac{1}{-\infty} = 0$$

The straight line that represents the oblique asymptote has equation:

$$y = 2x$$

where $q = 0$, the straight line intersects the origin of the axes.

11.2.8.4. Maximum and/or Minimum Points

We verify if the function has maximum or minimum points, checking if the first derivative is equal to 0 at some point and if for these points the second derivative is > 0 or < 0.

$$f'(x_i) = 0$$

$$f''(x_i) > 0 \quad or \quad f''(x_i) < 0$$

First derivative:

$$f'(x) = \frac{d}{dx}\left(x - \sqrt{x^2-1}\right) = \frac{d}{dx}\left(x - \left(x^2-1\right)^{\frac{1}{2}}\right) =$$

$$= 1 - \frac{1}{2}\left(x^2-1\right)^{-\frac{1}{2}} \cdot 2x = 1 - x\left(x^2-1\right)^{-\frac{1}{2}}$$

By assigning:

$$f'(x) = 1 - x\left(x^2-1\right)^{-\frac{1}{2}} = 0$$

it should be:

$$1 - x\left(x^2-1\right)^{-\frac{1}{2}} = 0 \to x\left(x^2-1\right)^{-\frac{1}{2}} = 1 \to x = \left(x^2-1\right)^{\frac{1}{2}} \to x^2 = x^2 - 1$$

The equation does not admit a solution, so there are no local points of maximum or minimum.

11.2.8.5. Inflection Points

We verify whether the function has inflection points, checking whether the second derivative is equal to 0.

$$f''(x)=\frac{d}{dx}\left[1-x\left(x^2-1\right)^{-\frac{1}{2}}\right]=-\left(x^2-1\right)^{-\frac{1}{2}}-x\left(-\frac{1}{2}\right)\left(x^2-1\right)^{-\frac{3}{2}}\cdot 2x=$$

$$=-\frac{1}{\left(x^2-1\right)^{\frac{1}{2}}}+\frac{x^2}{\left(x^2-1\right)^{\frac{3}{2}}}=\frac{-\left(x^2-1\right)+x^2}{\left(x^2-1\right)^{\frac{3}{2}}}=\frac{1}{\left(x^2-1\right)^{\frac{3}{2}}}>0$$

The second derivative of the function is never equal to 0, so there are no inflection points.

As it regards the curvature of the function, being:

$$f''(x)=\frac{1}{\left(x^2-1\right)^{\frac{3}{2}}}>0 \text{ (for } x<-1 \text{ and for } x>1)$$

The function is convex on the two definition intervals.

11.2.8.6. Additional Considerations on the Maximum Points

While there are no relative maximum points, it can be observed that, by identifying the domain with A:

$$A=\left]-\infty,-1\right]\cup\left[1,+\infty\right[$$

the function has an absolute maximum when it is superiorly bounded and the limit superior belongs to the codomain:

$$\exists \sup \mathcal{C}(f)\in\mathcal{C}(f) \qquad \Rightarrow \qquad \sup\mathcal{C}(f)=\max\mathcal{C}(f)$$

$$\max f=\sup_A f\equiv\sup\mathcal{C}(f)$$

In the present case:

$$\sup_A f=f(1)=1\equiv\sup\mathcal{C}(f)$$

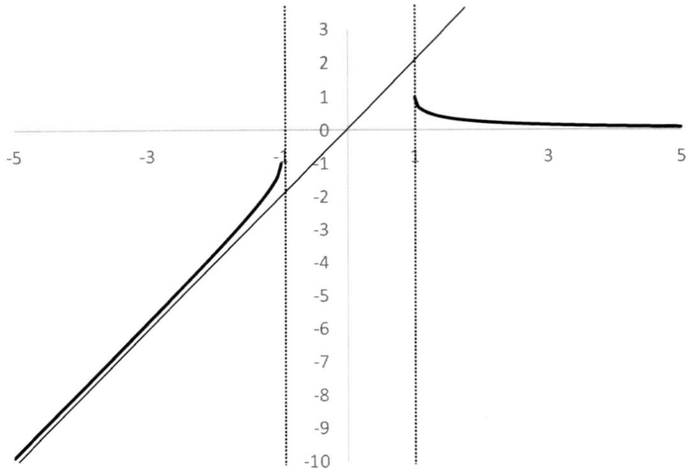

Figure 11.8. Graph of the function $f(x) = e^{-x^2}$.

For which:

$$\max f = 1$$

$x = 1$ is an absolute maximum point of the function.
The graph of the function is shown in Figure 11.8.

11.2.9. Problem 11.9

Study the function:

$$f(x) = xe^{-x}$$

11.2.9.1. Domain of the Function

The domain of the function is ***R***:

$$f : \boldsymbol{R} \to \boldsymbol{R}$$

11.2.9.2. Intersection with the Axes

We verify whether the function intersects the x, y axes:

$$x = 0 \quad \to \quad f(0) = ?$$
$$y = f(x) = 0 \quad \to \quad x = ?$$

In the present case, for $x = 0$:

$$y = f(0) = 0 \cdot e^{0} = 0$$

The function intersects the y-axis at the point $y = 0$.
For $y = 0$:

$$y = f(x) = x \cdot e^{-x} = 0 \quad \rightarrow \quad x = 0$$

The function intersects the x-axis at the point $x = 0$, so the function intersects the axes at the origin.

11.2.9.3. Asymptotes

There is a vertical asymptote if:

$$\lim_{x \to x_0 +} xe^{-x} = \pm\infty$$

$$\lim_{x \to x_0 -} xe^{-x} = \pm\infty$$

The function is continuous, so:

$$\lim_{x \to x_0} xe^{-x} = f(x_0) = x_0 e^{-x_0} \ \left(x_0 \in \boldsymbol{R},\ x_0 e^{-x_0} \neq \pm\infty \right)$$

The function has no vertical asymptotes, since $+\infty$ and $-\infty$ do not belong to $\boldsymbol{R}$, so it cannot be $f(x_0) = \pm\infty$.

There is a horizontal asymptote if:

$$\lim_{x \to -\infty} xe^{-x} = l_1,\ l_1 \in \boldsymbol{R}$$

$$\lim_{x \to +\infty} xe^{-x} = l_2,\ l_2 \in \boldsymbol{R}$$

We have:

$$\lim_{x \to +\infty} xe^{-x} = +\infty \cdot 0$$

it is an indeterminate form that can be written as::

$$\lim_{x\to+\infty} xe^{-x} = \lim_{x\to+\infty} \frac{x}{e^x} = \frac{+\infty}{+\infty}$$

which of course is still an indeterminate form.

We can apply De L'Hôpital's rule:

$$\lim_{x\to+\infty} \frac{x}{e^x} = \lim_{x\to+\infty} \frac{\frac{d}{dx}(x)}{\frac{d}{dx}(e^x)} = \lim_{x\to+\infty} \frac{1}{e^x} = \frac{1}{+\infty} = 0$$

For $x \to +\infty$ there is a horizontal asymptote at $y = 0$.

For $x \to -\infty$ the limit is:

$$\lim_{x\to-\infty} xe^{-x} = \lim_{x\to-\infty} \frac{x}{e^x} = \frac{-\infty}{0} = -\infty$$

There is no horizontal asymptote for $x \to -\infty$, so it is necessary to check whether there is an oblique asymptote:

$$\lim_{x\to-\infty} \frac{xe^{-x}}{x} = m$$

We have:

$$\lim_{x\to-\infty} \frac{xe^{-x}}{x} = \lim_{x\to-\infty} e^{-x} = \lim_{x\to-\infty} \frac{1}{e^x} = \frac{1}{0} = +\infty$$

There is no oblique asymptote for $x \to -\infty$.

11.2.9.4. Maximum and/or Minimum Points

We verify if the function has maximum or minimum points, checking if the first derivative is equal to 0 at some point and if for these points the second derivative is > 0 or < 0.

$$f'(x_i) = 0$$

$$f''(x_i) > 0 \quad or \quad f''(x_i) < 0$$

First derivative:

$$f'(x) = \frac{d}{dx}\left(xe^{-x}\right) = e^{-x} + x\left(-e^{-x}\right) = e^{-x}(1-x)$$

$$f'(x) = e^{-x}(1-x) = 0 \quad \rightarrow \quad 1-x=0 \quad \rightarrow \quad x=1$$

The first derivative is equal to 0 at the point $x = 1$.
Second derivative:

$$f''(x) = \frac{d}{dx}\left(e^{-x}(1-x)\right) = -e^{-x}(1-x) + e^{-x}(-1) =$$
$$= -e^{-x} + xe^{-x} - e^{-x} = e^{-x}(x-2)$$

The second derivative, at the points where the first derivative is equal to 0, it is:

$$f''(1) = e^{-1}(1-2) = -e^{-1} < 0$$

$x = 1$ is a point of maximum.
The value of the function at the maximum point is:

$$f(1) = 1 \cdot e^{-1} = e^{-1} = \frac{1}{e}$$

11.2.9.5. Inflection Points

We verify whether the function has inflection points, checking whether the second derivative is equal to 0.

$$f''(x) = e^{-x}(x-2) = 0$$
$$f''(x) = 0 \quad \Leftrightarrow \quad (x-2) = 0 \quad \rightarrow \quad x = 2$$

The function has one inflection point at the value $x = 2$.
The value of the function at the inflection point is:

$$f(2) = xe^{-x} = 2 \cdot e^{-2} = \frac{2}{e^2}$$

It can be observed that:

$$f(2) < f(1)$$

In fact, being $2 < e$:

$$\frac{2}{e^2} < \frac{e}{e^2} \equiv \frac{1}{e}$$

The graph of the function is shown in Figure 11.9.

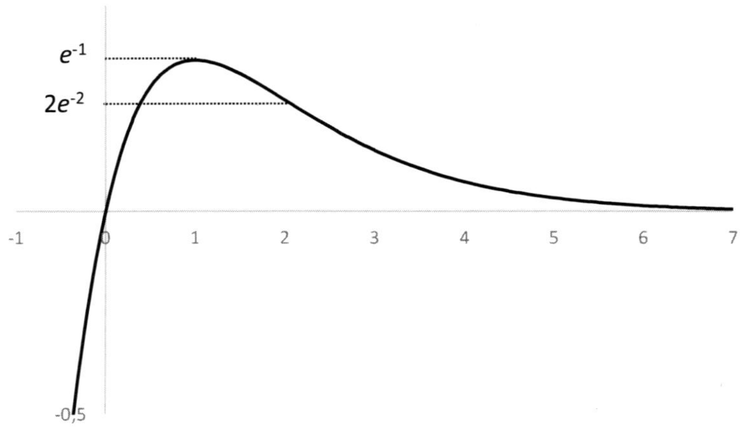

Figure 11.9. Graph of the function $f(x) = xe^{-x}$.

11.2.10. Problem 11.10

Study the function:

$$f(x) = (x-1)\log(x-1)$$

11.2.10.1. Domain of the Function

The domain of the function is determined by the logarithm, defined for values of the argument> 0:

$$x - 1 > 0 \quad \rightarrow \quad x > 1$$

The function is therefore defined by:

$$f :]1, +\infty[\rightarrow \boldsymbol{R}$$

11.2.10.2. Intersection with the Axes

We verify whether the function intersects the *x*, *y* axes:

$$x = 0 \quad \rightarrow \quad f(0) = ?$$
$$y = f(x) = 0 \quad \rightarrow \quad x = ?$$

The function is not defined for $x = 0$, so it does not intersect the *y*-axis.
For the intersection with the *x*-axis, it should be $y = 0$:

$$y = f(x) = (x-1)\log(x-1) = 0$$

Given the domain, the only condition for which the function is equal to 0 it is:

$$\log(x-1) = 0$$

By applying the exponential function, i.e., the inverse function of logarithm:

$$e^{\log(x-1)} = e^0 \quad \rightarrow \quad x - 1 = 1 \quad \rightarrow \quad x = 2$$

The function intersects the *x*-axis at the point $x = 2$.

11.2.10.3. Asymptotes

We have seen that the function does not intersect the *y*-axis, so we calculate the limit:

$$\lim_{x \to 1+} (x-1)\log(x-1) = 0 \cdot (-\infty)$$

It is an indeterminate form, we can apply De L'Hôpital's rule:

$$\lim_{x \to 1+} (x-1)\log(x-1) = \lim_{x \to 1+} \frac{\log(x-1)}{(x-1)^{-1}} = \lim_{x \to 1+} \frac{\frac{d}{dx}(\log(x-1))}{\frac{d}{dx}\left((x-1)^{-1}\right)} =$$

$$= \lim_{x \to 1+} \frac{\frac{1}{x-1}}{-(x-1)^{-2}} = \lim_{x \to 1+} \left[-\frac{1}{(x-1)^{-1}} \right] = \lim_{x \to 1+} [-(x-1)] = 0$$

The function tends to 0 for $x \to 1+$, so there is no vertical asymptote.

Let's check if there is a horizontal asymptote. We have:

$$\lim_{x\to+\infty}(x-1)\log(x-1)=+\infty\cdot(+\infty)=+\infty$$

The function tends to $+\infty$ for $x \to +\infty$. There is no horizontal asymptote, so it is necessary to check whether there is an oblique asymptote:

$$\lim_{x\to+\infty}\frac{f(x)}{x}=m$$

In this case:

$$\lim_{x\to+\infty}\frac{(x-1)\log(x-1)}{x}=\lim_{x\to+\infty}\frac{x-1}{x}\cdot\log(x-1)=$$
$$=\lim_{x\to+\infty}\left(1-\frac{1}{x}\right)\cdot\log(x-1)=(1-0)\cdot(+\infty)=+\infty$$

There is no oblique asymptote.

11.2.10.4. Maximum and/or Minimum Points

We verify if the function has maximum or minimum points, checking if the first derivative is equal to 0 at some point and if for these points the second derivative is > 0 or < 0.

$$f'(x_i)=0$$
$$f''(x_i)>0 \qquad or \qquad f''(x_i)<0$$

First derivative:

$$f'(x)=\frac{d}{dx}[(x-1)\log(x-1)]=\log(x-1)+\frac{x-1}{x-1}=\log(x-1)+1$$

$$f'(x)=\log(x-1)+1=0 \quad\to\quad \log(x-1)=-1$$

By applying the exponential function we find the value in which the first derivative is = 0:

$$e^{\log(x-1)} = e^{-1} \quad \rightarrow \quad x-1 = e^{-1} \quad \rightarrow \quad x = 1+e^{-1}$$

Second derivative:

$$f''(x) = \frac{d}{dx}\left[\log(x-1)+1\right] = \frac{1}{x-1}$$

The second derivative, at the point where the first derivative is equal to 0, it is:

$$f''\left(1+e^{-1}\right) = \frac{1}{1+e^{-1}-1} = e > 0$$

$x = 1+e^{-1}$ is a point of minimum.
The value of the function at the minimum point is:

$$f\left(1+e^{-1}\right) = \left(1+e^{-1}-1\right)\log\left(1+e^{-1}-1\right) =$$
$$= e^{-1}\log\left(e^{-1}\right) = e^{-1}\left(-\log e\right) = -e^{-1}$$

11.2.10.5. Inflection Points

We verify whether the function has inflection points, checking whether the second derivative is equal to 0.

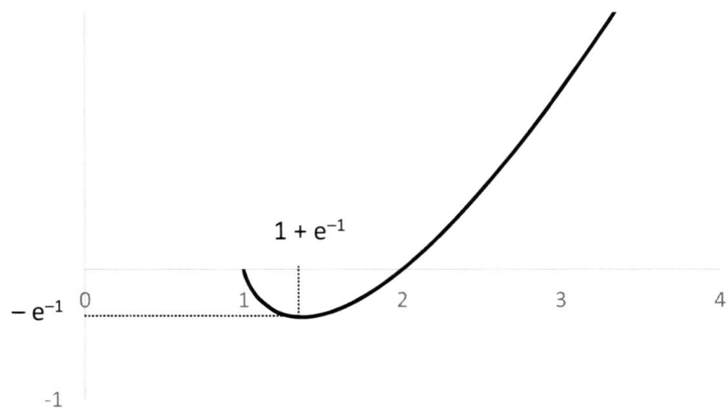

Figure 11.10. Graph of the function $f(x) = (x-1)\log(x-1)$.

$$f''(x) = 0$$

$\dfrac{1}{x-1} = 0$ it has no solution for $x \in \boldsymbol{R}$

The second derivative is not equal to 0, so there are no inflection points.

The graph of the function is shown in Figure 11.10.

11.2.11. Problem 11.11

Study the function:

$$f(x) = \frac{\log x}{x^3}$$

11.2.11.1. Domain of the Function

The domain of the function is determined by $\log x$ to the numerator, since the condition $x \neq 0$ relating to the denominator is already included in the domain of the logarithm function.

The function is therefore defined by:

$$f : \boldsymbol{R}^+ \to \boldsymbol{R}$$

11.2.11.2. Intersection with the Axes

We verify whether the function intersects the x, y axes:

$$x = 0 \quad \to \quad f(0) = ?$$

$$y = f(x) = 0 \quad \to \quad x = ?$$

The function is not defined for $x = 0$, so it does not intersect the y-axis.

For $y = 0$:

$$y = f(x) = \frac{\log x}{x^3} = 0$$

The only solution is that for which the numerator is equal to 0:

$$\log x = 0 \quad \Rightarrow \quad x = 1$$

The function intersects the x-axis at the point $x = 1$.

11.2.11.3. Asymptotes

There is a vertical asymptote if:

$$\lim_{x \to x_0 +} \frac{\log x}{x^3} = \pm\infty$$

$$\lim_{x \to x_0 -} \frac{\log x}{x^3} = \pm\infty$$

Given the function definition domain, we can check the limit for $x \to 0^+$.

$$\lim_{x \to 0+} \frac{\log x}{x^3} = \frac{-\infty}{0^+} = -\infty$$

There is a vertical asymptote for $x \to 0^+$.
There is a horizontal asymptote if:

$$\lim_{x \to +\infty} \frac{\log x}{x^3} = l,\ l \in \boldsymbol{R}$$

In the present case:

$$\lim_{x \to +\infty} \frac{\log x}{x^3} = \frac{+\infty}{+\infty}$$

It is an indeterminate form, we can apply De L'Hôpital's rule:

$$\lim_{x \to +\infty} \frac{\log x}{x^3} = \lim_{x \to +\infty} \frac{\frac{d}{dx}(\log x)}{\frac{d}{dx}(x^3)} = \lim_{x \to +\infty} \frac{\frac{1}{x}}{3x^2} = \lim_{x \to +\infty} \frac{1}{3x^3} = \frac{1}{+\infty} = 0$$

For $x \to +\infty$ there is a horizontal asymptote at $y = 0$. Given the function definition domain and the existence of a horizontal asymptote, there can be no oblique asymptote.

11.2.11.4. Maximum and/or Minimum Points

We verify if the function has maximum or minimum points, checking if the first derivative is equal to 0 at some point and if for these points the second derivative is > 0 or < 0.

$$f'(x_i) = 0$$
$$f''(x_i) > 0 \qquad or \qquad f''(x_i) < 0$$

First derivative:

$$f'(x) = \frac{d}{dx}\left(\frac{\log x}{x^3}\right) = \frac{\frac{1}{x}\cdot x^3 - \log x \cdot 3x^2}{x^6} = \frac{x^2 - 3x^2 \cdot \log x}{x^6} =$$
$$= \frac{x^2(1 - 3\log x)}{x^6} = \frac{1 - 3\log x}{x^4} = 0$$

$$f'(x) = \frac{1 - 3\log x}{x^4} = 0$$

The first derivative is equal to 0 when the numerator is equal to 0:

$$1 - 3\log x = 0 \quad \Rightarrow \quad 3\log x = 1 \quad \Rightarrow \quad \log x = \frac{1}{3}$$

By applying the inverse function of logarithm, we find the point where the first derivative is = 0:

$$e^{\log x} = e^{\frac{1}{3}} \quad \Rightarrow \quad x = e^{\frac{1}{3}}$$

Second derivative:

$$f''(x) = \frac{d}{dx}\left(\frac{1 - 3\log x}{x^4}\right) = \frac{-\frac{3}{x}\cdot x^4 - (1 - 3\log x)\cdot 4x^3}{x^8} =$$
$$\frac{-3x^3 - (1 - 3\log x)\cdot 4x^3}{x^8} = \frac{-x^3(3 + 4 - 12\log x)}{x^8} =$$
$$= \frac{-(7 - 12\log x)}{x^5} = \frac{12\log x - 7}{x^5}$$

The second derivative, at the point where the first derivative is equal to 0, it is:

$$f''\left(e^{\frac{1}{3}}\right)=\frac{12\log e^{\frac{1}{3}}-7}{\left(e^{\frac{1}{3}}\right)^{5}}=\frac{12\cdot\frac{1}{3}-7}{e^{\frac{5}{3}}}=\frac{4-7}{e^{\frac{5}{3}}}=\frac{-3}{e^{\frac{5}{3}}}<0$$

$x=e^{\frac{1}{3}}$ is a point of maximum

The value of the function at the maximum point is:

$$f\left(e^{\frac{1}{3}}\right)=\frac{\log e^{\frac{1}{3}}}{\left(e^{\frac{1}{3}}\right)^{3}}=\frac{\frac{1}{3}}{e^{\frac{3}{3}}}=\frac{1}{3e}=\frac{1}{3}e^{-1}$$

11.2.11.5. Inflection Points

We verify whether the function has inflection points, checking whether the second derivative is equal to 0.

$$f''(x)=\frac{12\log x-7}{x^{5}}=0$$

The second derivative is equal to 0 when the numerator is equal to 0:

$$12\log x-7=0 \quad\Rightarrow\quad 12\log x=7 \quad\Rightarrow\quad \log x=\frac{7}{12}$$

By applying the inverse function of logarithm, we find:

$$e^{\log x}=e^{\frac{7}{12}} \quad\Rightarrow\quad x=e^{\frac{7}{12}}$$

The function has one inflection point at the value $x=e^{\frac{7}{12}}$

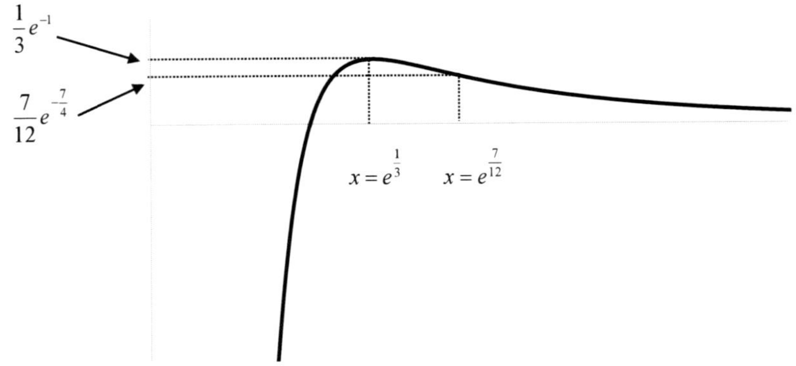

Figure 11.11. Graph of the function $f(x) = \dfrac{\log x}{x^3}$.

The value of the function at the inflection point is:

$$f\left(e^{\frac{7}{12}}\right) = \frac{\log e^{\frac{7}{12}}}{\left(e^{\frac{7}{12}}\right)^3} = \frac{\frac{7}{12}}{e^{\frac{7}{4}}} = \frac{7}{12e^{\frac{7}{4}}} = \frac{7}{12}e^{-\frac{7}{4}}$$

The graph of the function is shown in Figure 11.11.

11.1.12. Problem 11.12

Study the function:

$$f(x) = \frac{x}{x^2 - 1}$$

11.2.12.1. Domain of the Function

The domain of the function is determined by the values for which:

$$x^2 - 1 \neq 0 \quad \rightarrow \quad x^2 \neq 1 \quad \rightarrow \quad |x| \neq 1$$

The function is therefore defined by:

$$f : \boldsymbol{R} - \{-1,1\} \rightarrow \boldsymbol{R}$$

11.2.12.2. Intersection with the Axes

We verify whether the function intersects the x, y axes:

$$x = 0 \quad \rightarrow \quad f(0) = ?$$

$$y = f(x) = 0 \quad \rightarrow \quad x = ?$$

For $x = 0$:

$$y = f(0) = \frac{0}{0-1} = 0$$

The function intersects the y-axis at the point $y = 0$.
For $y = 0$:

$$y = f(x) = \frac{x}{x^2 - 1} = 0$$

The only solution is that for which the numerator is equal to 0:

$$x = 0$$

The function intersects the x-axis at point $x = 0$, so the function intersects the axes at the origin (0,0).

11.2.12.3. Asymptotes

There is a vertical asymptote if:

$$\lim_{x \to x_0 +} \frac{x}{x^2 - 1} = \pm\infty$$

$$\lim_{x \to x_0 -} \frac{x}{x^2 - 1} = \pm\infty$$

Given the domain of the function, let's check the limits for $x_0 = -1$ and $x_0 = +1$.

$$\lim_{x \to -1+} \frac{x}{x^2 - 1} = \frac{-1}{0^-} = +\infty$$

the limit for x → –1+ is calculated for $-1 < x < 0$, that is $|x| < 1$, from which 0^- to the denominator.

$$\lim_{x \to -1-} \frac{x}{x^2 - 1} = \frac{-1}{0^+} = -\infty$$

the limit for x → -1^- is calculated for $x < -1$, that is $|x| > 1$, from which 0^+ to the denominator.

$$\lim_{x \to +1-} \frac{x}{x^2 - 1} = \frac{1}{0^-} = -\infty$$

the limit for x → $+1^-$ is calculated for $0 < x < 1$, that is $|x| < 1$, from which 0^- to the denominator.

$$\lim_{x \to +1+} \frac{x}{x^2 - 1} = \frac{1}{0^+} = +\infty$$

the limit for x → -1^+ is calculated for $x > 1$, that is $|x| > 1$, from which 0^+ to the denominator.

There are two vertical asymptotes with four infinite branches at the points $x_0 = -1$ and $x_0 = +1$.

There is a horizontal asymptote if:

$$\lim_{x \to -\infty} \frac{x}{x^2 - 1} = l_1,\ l_1 \in \boldsymbol{R}$$

$$\lim_{x \to +\infty} \frac{x}{x^2 - 1} = l_2,\ l_2 \in \boldsymbol{R}$$

For the function it holds:

$$\lim_{x \to +\infty} \frac{x}{x^2 - 1} = \frac{+\infty}{+\infty}$$

It is an indeterminate form, which can be resolved in the following way:

$$\lim_{x \to +\infty} \frac{x}{x^2 - 1} = \lim_{x \to +\infty} \frac{1}{\frac{x^2 - 1}{x}} = \lim_{x \to +\infty} \frac{1}{x - \frac{1}{x}} = \frac{1}{\infty - \frac{1}{\infty}} = 0$$

Alternatively, the limit can be calculated by applying the De L'Hôpital's rule:

$$\lim_{x \to +\infty} \frac{x}{x^2 - 1} = \lim_{x \to +\infty} \frac{\frac{d}{dx}x}{\frac{d}{dx}\left(x^2 - 1\right)} = \lim_{x \to +\infty} \frac{1}{2x} = 0$$

For $x \to +\infty$ there is a horizontal asymptote at $y = 0$.
For $x \to -\infty$ it holds:

$$\lim_{x \to -\infty} \frac{x}{x^2 - 1} = \frac{-\infty}{+\infty}$$

It is an indeterminate form, we can apply De L'Hôpital's rule:

$$\lim_{x \to -\infty} \frac{x}{x^2 - 1} = \lim_{x \to -\infty} \frac{\frac{d}{dx}x}{\frac{d}{dx}\left(x^2 - 1\right)} = \lim_{x \to -\infty} \frac{1}{2x} = 0$$

For $x \to -\infty$ there is a horizontal asymptote at $y = 0$. Given the existence of two horizontal asymptotes, there can be no oblique asymptotes.

11.2.12.4. Maximum and/or Minimum Points

We verify if the function has maximum or minimum points, checking if the first derivative is equal to 0 at some point and if for these points the second derivative is > 0 or < 0.

$$f'(x_i) = 0$$

$$f''(x_i) > 0 \quad or \quad f''(x_i) < 0$$

First derivative:

$$f'(x)=\frac{d}{dx}\left(\frac{x}{x^2-1}\right)=\frac{1\cdot(x^2-1)-x\cdot 2x}{(x^2-1)^2}=\frac{x^2-1-2x^2}{(x^2-1)^2}=-\frac{x^2+1}{(x^2-1)^2}$$

$$f'(x)=-\frac{x^2+1}{(x^2-1)^2}=0$$

The equation does not admit a solution, so there are no local points of maximum or minimum.

11.2.12.5. Inflection Points

We verify whether the function has inflection points, checking whether the second derivative is equal to 0.

$$f''(x)=\frac{d}{dx}\left[-\frac{x^2+1}{(x^2-1)^2}\right]=-\frac{2x\cdot(x^2-1)^2-(x^2+1)\cdot 2(x^2-1)\cdot 2x}{(x^2-1)^4}=$$

$$=-\frac{2x\cdot(x^2-1)\cdot[(x^2-1)-2(x^2+1)]}{(x^2-1)^4}=-\frac{2x\cdot(x^2-1)\cdot[x^2-1-2x^2-2]}{(x^2-1)^4}=$$

$$=-\frac{2x\cdot(x^2-1)\cdot[-x^2-3]}{(x^2-1)^4}=\frac{2x\cdot(x^2-1)\cdot(x^2+3)}{(x^2-1)^4}=\frac{2x\cdot(x^2+3)}{(x^2-1)^3}$$

$$f''(x)=\frac{2x\cdot(x^2+3)}{(x^2-1)^3}=0 \quad\Rightarrow\quad x=0$$

The function has one inflection point at the value $x = 0$.

As regards the curvature of the function, it holds:

$$f''(x)=\frac{2x\cdot(x^2+3)}{(x^2-1)^3}>0$$

for $-1 < x < 0$, the function is convex

$$f''(x)=\frac{2x\cdot(x^2+3)}{(x^2-1)^3}<0$$

for $0 < x < +1$, the function is concave

The value of the function at the inflection point is:

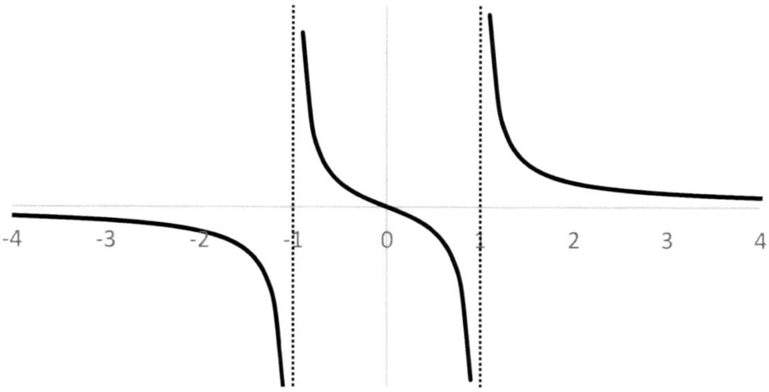

Figure 11.12. Graph of the function $f(x) = \frac{x}{x^2 - 1}$.

$$y = f(0) = \frac{0}{0-1} = 0$$

The graph of the function is shown in Figure 11.12.

11.1.13. Problem 11.13

Study the function:

$$f(x) = \frac{\log(2x+5)}{2x+5}$$

11.2.13.1. Domain of the Function

The domain of the function is given by the intersection of the function domains in the numerator and the denominator, minus the values in which the denominator is equal to 0.

The logarithm function is defined by:

$$\log : \boldsymbol{R}^+ \rightarrow \boldsymbol{R}$$

In this case, given the argument of the logarithm function we have::

$$2x + 5 > 0 \quad \Rightarrow \quad 2x > -5 \quad \Rightarrow \quad x > -\frac{5}{2}$$

so the domain of the numerator is:

$$D = \left]-\frac{5}{2}, +\infty\right[$$

The denominator (its domain is $\boldsymbol{R}$) equals to 0 for:

$$2x + 5 = 0 \quad \Rightarrow \quad x = -\frac{5}{2}$$

value already excluded from the domain of the numerator.

The function is therefore defined by:

$$f : \left]-\frac{5}{2}, +\infty\right[\to \boldsymbol{R}$$

11.2.13.2. Intersection with the Axes

We verify whether the function intersects the x, y axes:

$$x = 0 \quad \rightarrow \quad f(0) = ?$$

$$y = f(x) = 0 \quad \rightarrow \quad x = ?$$

For $x = 0$:

$$y = f(0) = \frac{\log(2 \cdot 0 + 5)}{2 \cdot 0 + 5} = \frac{\log 5}{5}$$

The function intersects the y-axis at the point $y = \dfrac{\log 5}{5}$

For $y = 0$:

$$y = f(x) = \frac{\log(2x + 5)}{2x + 5} = 0$$

The only solution is that for which the numerator is equal to 0:

$$\log(2x + 5) = 0$$

We apply the inverse function of logarithm:

$$e^{\log(2x+5)} = e^0 \quad \Rightarrow \quad 2x+5=1 \quad \Rightarrow \quad x=\frac{1-5}{2}=-2$$

The function intersects the x-axis at the point $x=-2$, so the function intersects the axes at the points $\left(0,\frac{\log 5}{5}\right)$ and $(-2,0)$.

11.2.13.3. Asymptotes

There is a vertical asymptote if:

$$\lim_{x\to x_0+}\frac{\log(2x+5)}{2x+5}=\pm\infty$$
$$\lim_{x\to x_0-}\frac{\log(2x+5)}{2x+5}=\pm\infty$$

Given the function definition domain, we can check the limit for $x_0^+=-\frac{5^+}{2}$:

$$\lim_{x\to -\frac{5}{2}+}\frac{\log(2x+5)}{2x+5}=\frac{\log(-5^+ +5)}{-5^+ +5}=\frac{-\infty}{0^+}=-\infty$$

There is a vertical asymptote at $x_0=-\frac{5}{2}$ with one infinite branch to $-\infty$.

Given the function definition domain, for horizontal asymptotes we check the limit only for $x\to+\infty$.

$$\lim_{x\to+\infty}\frac{\log(2x+5)}{2x+5}=l_1,\ l_1\in\boldsymbol{R}$$

We have:

$$\lim_{x\to+\infty}\frac{\log(2x+5)}{2x+5}=\frac{+\infty}{+\infty}$$

It is an indeterminate form, we can apply De L'Hôpital's rule:

$$\lim_{x\to+\infty}\frac{\log(2x+5)}{2x+5}=\lim_{x\to+\infty}\frac{\frac{d}{dx}\log(2x+5)}{\frac{d}{dx}(2x+5)}=$$

$$=\lim_{x\to+\infty}\frac{\frac{2}{2x+5}}{2}=\lim_{x\to+\infty}\frac{1}{2x+5}=\frac{1}{+\infty}=0$$

For $x \to +\infty$ there is a horizontal asymptote at $y = 0$.

11.2.13.4. Maximum and/or Minimum Points

We verify if the function has maximum or minimum points, checking if the first derivative is equal to 0 at some point and if for these points the second derivative is > 0 or < 0.

$$f'(x_i)=0$$
$$f''(x_i)>0 \quad or \quad f''(x_i)<0$$

First derivative:

$$f'(x)=\frac{d}{dx}\left(\frac{\log(2x+5)}{2x+5}\right)=\frac{\frac{2}{2x+5}\cdot(2x+5)-\log(2x+5)\cdot 2}{(2x+5)^2}=$$
$$=\frac{2-2\log(2x+5)}{(2x+5)^2}$$

$$f'(x)=\frac{2-2\log(2x+5)}{(2x+5)^2}=0$$

The only solution is that for which the numerator is equal to 0:

$$2-2\log(2x+5)=0 \quad\Rightarrow\quad \log(2x+5)=1 \quad\Rightarrow$$
$$e^{\log(2x+5)}=e^1 \quad\Rightarrow\quad 2x+5=e$$

The first derivative is equal to 0 at the point:

$$x = \frac{e-5}{2}$$

Second derivative:

$$f''(x) = \frac{d}{dx}\left(\frac{2-2\log(2x+5)}{(2x+5)^2}\right) =$$

$$= \frac{-2\cdot\frac{2}{2x+5}\cdot(2x+5)^2 - [2-2\log(2x+5)]\cdot 2(2x+5)\cdot 2}{(2x+5)^4} =$$

$$= \frac{-4(2x+5)-4(2x+5)\cdot[2-2\log(2x+5)]}{(2x+5)^4} =$$

$$= \frac{-4\cdot[1+2-2\log(2x+5)]}{(2x+5)^3} = \frac{4\cdot[2\log(2x+5)-3]}{(2x+5)^3}$$

The second derivative, at the point x where the first derivative is equal to 0, it is:

$$f''\left(\frac{e-5}{2}\right) = \frac{4\cdot\left[2\log\left(2\cdot\frac{e-5}{2}+5\right)-3\right]}{\left(2\cdot\frac{e-5}{2}+5\right)^3} = \frac{4\cdot[2\log(e-5+5)-3]}{(e-5+5)^3} =$$

$$= \frac{4\cdot[2\log e-3]}{e^3} = \frac{4\cdot(2-3)}{e^3} < 0$$

$x = \frac{e-5}{2}$ is a point of maximum.

The value of the function at the maximum point is:

$$f\left(\frac{e-5}{2}\right) = \frac{\log\left(2\cdot\frac{e-5}{2}+5\right)}{2\cdot\frac{e-5}{2}+5} = \frac{\log e}{e} = e^{-1}$$

11.2.13.5. Inflection Points

We verify whether the function has inflection points, checking whether the second derivative is equal to 0.

$$f''(x) = \frac{4\cdot[2\log(2x+5)-3]}{(2x+5)^3} = 0$$

The only solution is that for which the numerator is equal to 0:

$$2\log(2x+5)-3=0 \quad \Rightarrow \quad \log(2x+5)=\frac{3}{2} \quad \Rightarrow$$

$$e^{\log(2x+5)}=e^{\frac{3}{2}} \quad \Rightarrow \quad 2x+5=e^{\frac{3}{2}}$$

The second derivative is equal to 0 at the point:

$$x=\frac{e^{\frac{3}{2}}-5}{2}$$

The value of the function at the inflection point is:

$$f\left(\frac{e^{\frac{3}{2}}-5}{2}\right)=\frac{\log\left(2\cdot\frac{e^{\frac{3}{2}}-5}{2}+5\right)}{2\cdot\frac{e^{\frac{3}{2}}-5}{2}+5}=\frac{\log e^{\frac{3}{2}}}{e^{\frac{3}{2}}}=\frac{\frac{3}{2}\log e}{e^{\frac{3}{2}}}=\frac{3}{2}e^{-\frac{3}{2}}$$

The graph of the function is shown in Figure 11.13.

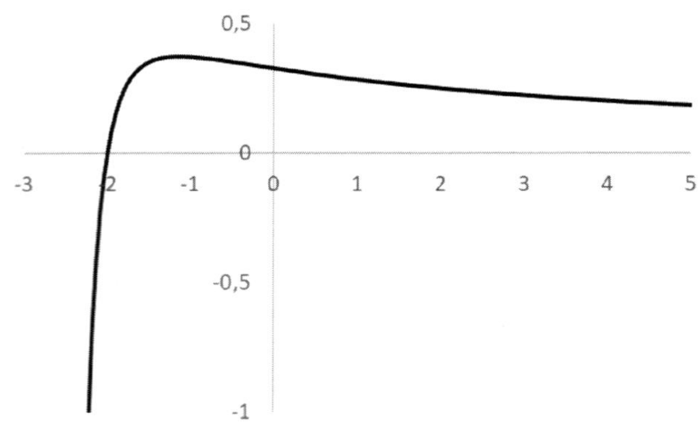

Figure 11.13. Graph of the function $f(x)=\dfrac{\log(2x+5)}{2x+5}$.

11.2.14. Problem 11.14

Study the function:

$$f(x) = \frac{1}{e^x + e^{-x}}$$

11.2.14.1. Domain of the Function

The domain of the function is given by the intersection of the function domains in the numerator and the denominator, minus the values in which the denominator is equal to 0.

In this case, the numerator is a constant so we consider only the domain of the denominator minus the values in which it is equal to 0.

The denominator is made up of the sum of two exponentials, so its domain is $\boldsymbol{R}$:

$$e^x + e^{-x} : \boldsymbol{R} \to \boldsymbol{R}^+$$

The exponential function is always > 0, so in this case the denominator is never equal to 0.

The function is therefore defined by:

$$f : \boldsymbol{R} \to \boldsymbol{R}^+$$

11.2.14.2. Intersection with the Axes

We verify whether the function intersects the x, y axes:

$$x = 0 \quad \to \quad f(0) = ?$$
$$y = f(x) = 0 \quad \to \quad x = ?$$

For $x = 0$:

$$y = f(0) = \frac{1}{e^0 + e^{-0}} = \frac{1}{1+1} = \frac{1}{2}$$

The function intersects the y-axis at the point $y = \frac{1}{2}$

For $y = 0$:

$$y = f(x) = \frac{1}{e^x + e^{-x}} = 0$$

The numerator is a constant ($1 \neq 0$), so the equation does not admit a solution and the function does not intersect the x axis.

11.2.14.3. Asymptotes

There is a vertical asymptote if:

$$\lim_{x \to x_0+} \frac{1}{e^x + e^{-x}} = \pm\infty$$

$$\lim_{x \to x_0-} \frac{1}{e^x + e^{-x}} = \pm\infty$$

The function is continuous on $\boldsymbol{R}$, so there are no discontinuity points x_0 and the function has no vertical asymptotes.

There is a horizontal asymptote if:

$$\lim_{x \to +\infty} \frac{1}{e^x + e^{-x}} = l_1,\ l_1 \in \boldsymbol{R}$$

$$\lim_{x \to -\infty} \frac{1}{e^x + e^{-x}} = l_2,\ l_2 \in \boldsymbol{R}$$

We have:

$$\lim_{x \to +\infty} \frac{1}{e^x + e^{-x}} = \frac{1}{e^{+\infty} + e^{-\infty}} = \frac{1}{+\infty + 0} = \frac{1}{+\infty} = 0$$

$$\lim_{x \to -\infty} \frac{1}{e^x + e^{-x}} = \frac{1}{e^{-\infty} + e^{+\infty}} = \frac{1}{0 + \infty} = \frac{1}{+\infty} = 0$$

There is a horizontal asymptote at $y = 0$ with two infinite branches for $x \to +\infty$ and $x \to -\infty$.

11.2.14.4. Maximum and/or Minimum Points

We verify if the function has maximum or minimum points, checking if the first derivative is equal to 0 at some point and if for these points the second derivative is > 0 or < 0.

$$f'(x_i) = 0$$

$$f''(x_i) > 0 \quad or \quad f''(x_i) < 0$$

First derivative:

$$f'(x) = \frac{d}{dx}\left(\frac{1}{e^x + e^{-x}}\right) = -\frac{\frac{d}{dx}\left(e^x + e^{-x}\right)}{\left(e^x + e^{-x}\right)^2} = -\frac{e^x - e^{-x}}{\left(e^x + e^{-x}\right)^2}$$

$$f'(x) = -\frac{e^x - e^{-x}}{\left(e^x + e^{-x}\right)^2} = 0$$

The only solution is that for which the numerator is equal to 0:

$$e^x - e^{-x} = 0 \quad \Rightarrow \quad e^x = e^{-x} \quad \Rightarrow \quad \log\left(e^x\right) = \log\left(e^{-x}\right) \Rightarrow$$

$$x = -x \quad \Leftrightarrow \quad x = 0$$

The first derivative is equal to 0 at the point $x = 0$.
Second derivative:

$$f''(x) = \frac{d}{dx}\left(-\frac{e^x - e^{-x}}{\left(e^x + e^{-x}\right)^2}\right) =$$

$$= -\frac{\left(e^x + e^{-x}\right)\cdot\left(e^x + e^{-x}\right)^2 - \left(e^x - e^{-x}\right)\cdot 2\left(e^x + e^{-x}\right)\cdot\left(e^x - e^{-x}\right)}{\left(e^x + e^{-x}\right)^4} =$$

$$= -\frac{\left(e^x + e^{-x}\right)\cdot\left[\left(e^x + e^{-x}\right)^2 - 2\left(e^x - e^{-x}\right)^2\right]}{\left(e^x + e^{-x}\right)^4} =$$

$$= -\frac{\left(e^x + e^{-x}\right)^2 - 2\left(e^x - e^{-x}\right)^2}{\left(e^x + e^{-x}\right)^3}$$

The second derivative, at the point x where the first derivative is equal to 0, it is:

$$f''(0) = -\frac{\left(e^0 + e^{-0}\right)^2 - 2\left(e^0 - e^{-0}\right)^2}{\left(e^0 + e^{-0}\right)^3} = -\frac{(1+1)^2 - 2(1-1)^2}{(1+1)^3} =$$

$$= -\frac{2^2}{2^3} = -\frac{1}{2} < 0$$

$x = 0$ is a point of maximum.
The value of the function at the maximum point is:

$$f(0) = \frac{1}{e^0 + e^{-0}} = \frac{1}{2}$$

11.2.14.5. Inflection Points

We verify whether the function has inflection points, checking whether the second derivative is equal to 0.

$$f''(x) = -\frac{(e^x + e^{-x})^2 - 2(e^x - e^{-x})^2}{(e^x + e^{-x})^3} = 0$$

The only solution is that for which the numerator is equal to 0:

$$(e^x + e^{-x})^2 - 2(e^x - e^{-x})^2 = 0$$

It can be observed that the numerator is formed by the difference between two squares, where $2 = (\sqrt{2})^2$, for which it holds:

$$a^2 - b^2 = (a+b)\cdot(a-b)$$

Thus:

$$(e^x + e^{-x})^2 - 2(e^x - e^{-x})^2 =$$
$$= \left[(e^x + e^{-x}) + \sqrt{2}(e^x - e^{-x})\right]\cdot\left[(e^x + e^{-x}) - \sqrt{2}(e^x - e^{-x})\right] = 0$$

The equation admits two solutions:

1) $(e^x + e^{-x}) + \sqrt{2}(e^x - e^{-x}) = 0$
2) $(e^x + e^{-x}) - \sqrt{2}(e^x - e^{-x}) = 0$

For the first solution, by collecting e^x and e^{-x}, we obtain:

$$\left(e^x+e^{-x}\right)+\sqrt{2}\left(e^x-e^{-x}\right)=0$$
$$\left(1+\sqrt{2}\right)e^x+\left(1-\sqrt{2}\right)e^{-x}=0$$
$$\left(1+\sqrt{2}\right)e^x=-\left(1-\sqrt{2}\right)e^{-x}$$
$$e^x\cdot e^x=-\frac{\left(1-\sqrt{2}\right)}{\left(1+\sqrt{2}\right)} \quad\Rightarrow\quad e^{2x}=\frac{-1+\sqrt{2}}{1+\sqrt{2}}=\frac{\sqrt{2}-1}{\sqrt{2}+1}$$
$$\log\left(e^{2x}\right)=\log\left(\frac{\sqrt{2}-1}{\sqrt{2}+1}\right) \quad\Rightarrow\quad 2x=\log\left(\frac{\sqrt{2}-1}{\sqrt{2}+1}\right)$$
$$x=\frac{1}{2}\log\left(\frac{\sqrt{2}-1}{\sqrt{2}+1}\right)$$

The value of the function at the inflection point is:

$$f\left(\frac{1}{2}\log\left(\frac{\sqrt{2}-1}{\sqrt{2}+1}\right)\right)=\frac{1}{e^{\frac{1}{2}\log\left(\frac{\sqrt{2}-1}{\sqrt{2}+1}\right)}+e^{-\frac{1}{2}\log\left(\frac{\sqrt{2}-1}{\sqrt{2}+1}\right)}}=$$
$$=\frac{1}{e^{\log\left(\frac{\sqrt{2}-1}{\sqrt{2}+1}\right)^{\frac{1}{2}}}+e^{\log\left(\frac{\sqrt{2}-1}{\sqrt{2}+1}\right)^{-\frac{1}{2}}}}=\frac{1}{\left(\frac{\sqrt{2}-1}{\sqrt{2}+1}\right)^{\frac{1}{2}}+\left(\frac{\sqrt{2}-1}{\sqrt{2}+1}\right)^{-\frac{1}{2}}}$$

Similarly, we obtain the second solution:

$$\left(e^x+e^{-x}\right)-\sqrt{2}\left(e^x-e^{-x}\right)=0$$
$$\left(1-\sqrt{2}\right)e^x+\left(1+\sqrt{2}\right)e^{-x}=0$$
$$\left(1-\sqrt{2}\right)e^x=-\left(1+\sqrt{2}\right)e^{-x}$$
$$e^x\cdot e^x=-\frac{\left(1+\sqrt{2}\right)}{\left(1-\sqrt{2}\right)} \quad\Rightarrow\quad e^{2x}=-\frac{1+\sqrt{2}}{1-\sqrt{2}}=\frac{1+\sqrt{2}}{\sqrt{2}-1}\equiv\frac{\sqrt{2}+1}{\sqrt{2}-1}$$
$$\log\left(e^{2x}\right)=\log\left(\frac{\sqrt{2}+1}{\sqrt{2}-1}\right) \quad\Rightarrow\quad 2x=\log\left(\frac{\sqrt{2}+1}{\sqrt{2}-1}\right)$$
$$x=\frac{1}{2}\log\left(\frac{\sqrt{2}+1}{\sqrt{2}-1}\right)$$

The value of the function at the inflection point is:

$$f\left(\frac{1}{2}\log\left(\frac{\sqrt{2}+1}{\sqrt{2}-1}\right)\right)=\frac{1}{e^{\frac{1}{2}\log\left(\frac{\sqrt{2}+1}{\sqrt{2}-1}\right)}+e^{-\frac{1}{2}\log\left(\frac{\sqrt{2}+1}{\sqrt{2}-1}\right)}}=$$

$$=\frac{1}{e^{\log\left(\frac{\sqrt{2}+1}{\sqrt{2}-1}\right)^{\frac{1}{2}}}+e^{\log\left(\frac{\sqrt{2}+1}{\sqrt{2}-1}\right)^{-\frac{1}{2}}}}=\frac{1}{\left(\frac{\sqrt{2}+1}{\sqrt{2}-1}\right)^{\frac{1}{2}}+\left(\frac{\sqrt{2}+1}{\sqrt{2}-1}\right)^{-\frac{1}{2}}}=$$

$$=\frac{1}{\left(\frac{\sqrt{2}-1}{\sqrt{2}+1}\right)^{-\frac{1}{2}}+\left(\frac{\sqrt{2}-1}{\sqrt{2}+1}\right)^{\frac{1}{2}}}$$

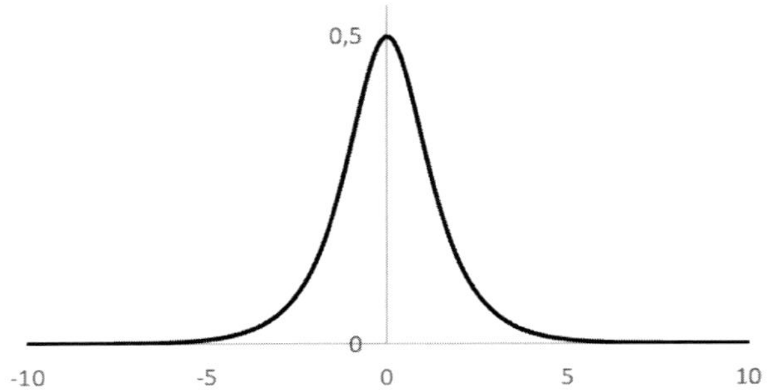

Figure 11.14. Graph of the function $f(x)=\dfrac{1}{e^{x}+e^{-x}}$.

The graph of the function is shown in Figure 11.14.

11.2.15. Problem 11.15

Study the function:

$$f(x)=\frac{x}{1-\sqrt{x}}$$

11.2.15.1. Domain of the Function

The function in the numerator is defined on ***R***, so the domain of the function depends on the denominator domain minus the values in which the denominator is equal to 0.

With regard to the square root, it is defined by:

$$\sqrt{x} \equiv x^{\frac{1}{2}}: \quad [0,+\infty[\to \boldsymbol{R}$$

The value for which the denominator is equal to 0 it is $x = 1$ which should be excluded. The function is therefore defined by:

$$f: [0,1[\cup]1,+\infty[\to \boldsymbol{R}$$

11.2.15.2. Intersection with the Axes

We verify whether the function intersects the x, y axes:

$$x = 0 \quad \to \quad f(0) = ?$$
$$y = f(x) = 0 \quad \to \quad x = ?$$

For $x = 0$:

$$y = f(0) = \frac{0}{1-\sqrt{0}} = \frac{0}{1} = 0$$

The function intersects the y-axis at the point $y = 0$

For $y = 0$:

$$y = f(x) = \frac{x}{1-\sqrt{x}} = 0 \quad \Rightarrow \quad x = 0$$

The function intersects the x-axis at point $x = 0$, so the function intersects the axes at the origin (0,0).

11.2.15.3. Asymptotes

There is a vertical asymptote if:

$$\lim_{x \to x_0+} \frac{x}{1-\sqrt{x}} = \pm\infty$$
$$\lim_{x \to x_0-} \frac{x}{1-\sqrt{x}} = \pm\infty$$

Given the domain of the function, let's check the limits for $x_0 = 1$:

$$\lim_{x\to 1+}\frac{x}{1-\sqrt{x}}=\frac{1}{0^-}=-\infty$$

the denominator tends to 0 from negative values;

$$\lim_{x\to 1-}\frac{x}{1-\sqrt{x}}=\frac{1}{0^+}=+\infty$$

the denominator tends to 0 from positive values.

There is a vertical asymptote at $x = 1$ with two infinite branches ($+\infty$ and $-\infty$).

Given the domain of the function, we check the limit only for $x \to +\infty$. There is a horizontal asymptote if

$$\lim_{x\to +\infty}\frac{x}{1-\sqrt{x}}=l_1,\ l_1\in \boldsymbol{R}$$

We have:

$$\lim_{x\to +\infty}\frac{x}{1-\sqrt{x}}=\frac{+\infty}{-\infty}$$

It is an indeterminate form, we can apply De L'Hôpital's rule:

$$\lim_{x\to +\infty}\frac{x}{1-\sqrt{x}}=\lim_{x\to +\infty}\frac{\frac{d}{dx}x}{\frac{d}{dx}\left(1-x^{\frac{1}{2}}\right)}=\lim_{x\to +\infty}\frac{1}{-\frac{1}{2}x^{-\frac{1}{2}}}=\lim_{x\to +\infty}-2x^{\frac{1}{2}}=-\infty$$

There is no horizontal asymptote for $x \to +\infty$, so it is necessary to check whether there is an oblique asymptote:

$$\lim_{x\to +\infty}\frac{f(x)}{x}=m,\ \lim_{x\to +\infty}\left[f(x)-mx\right]=q,\ m,q\in \boldsymbol{R}$$

For the coefficient m:

$$\lim_{x\to+\infty}\frac{f(x)}{x}=\lim_{x\to+\infty}\frac{1}{x}\frac{x}{1-\sqrt{x}}=\lim_{x\to+\infty}\frac{1}{1-\sqrt{x}}=\lim_{x\to+\infty}\frac{1}{-\infty}=0$$

Since m = 0, the function does not have an oblique asymptote. The result m = 0 is sufficient to establish that there is no oblique asymptote. In fact m = 0 corresponds to a horizontal straight line, that is a horizontal asymptote that, if there was, it should have been identified previously.

11.2.15.4. Maximum and/or Minimum Points

We verify if the function has maximum or minimum points, checking if the first derivative is equal to 0 at some point and if for these points the second derivative is > 0 or < 0.

$$f'(x_i)=0$$
$$f''(x_i)>0 \qquad or \qquad f''(x_i)<0$$

First derivative:

$$f'(x)=\frac{d}{dx}\left(\frac{x}{1-\sqrt{x}}\right)\equiv\frac{d}{dx}\left(\frac{x}{1-x^{\frac{1}{2}}}\right)=\frac{1\cdot\left(1-x^{\frac{1}{2}}\right)-x\left(-\frac{1}{2}x^{-\frac{1}{2}}\right)}{\left(1-x^{\frac{1}{2}}\right)^2}=$$

$$=\frac{1-x^{\frac{1}{2}}+\frac{1}{2}x^{\frac{1}{2}}}{\left(1-x^{\frac{1}{2}}\right)^2}=\frac{1-\frac{1}{2}x^{\frac{1}{2}}}{\left(1-x^{\frac{1}{2}}\right)^2}$$

$$f'(x)=\frac{1-\frac{1}{2}x^{\frac{1}{2}}}{\left(1-x^{\frac{1}{2}}\right)^2}=0$$

The only solution is that for which the numerator is equal to 0:

$$1-\frac{1}{2}x^{\frac{1}{2}}=0 \quad\Rightarrow\quad \frac{1}{2}x^{\frac{1}{2}}=1 \quad\Rightarrow\quad x^{\frac{1}{2}}=2 \quad\Rightarrow\quad x=4$$

The first derivative is equal to 0 at the point $x = 4$.
Second derivative:

$$f''(x) = \frac{d}{dx}\left[\frac{1-\frac{1}{2}x^{\frac{1}{2}}}{\left(1-x^{\frac{1}{2}}\right)^2}\right] =$$

$$= \frac{-\frac{1}{2}\cdot\frac{1}{2}x^{-\frac{1}{2}}\cdot\left(1-x^{\frac{1}{2}}\right)^2 - \left(1-\frac{1}{2}x^{\frac{1}{2}}\right)\cdot 2\left(1-x^{\frac{1}{2}}\right)\cdot\left(-\frac{1}{2}x^{-\frac{1}{2}}\right)}{\left(1-x^{\frac{1}{2}}\right)^4} =$$

$$= \frac{x^{-\frac{1}{2}}\cdot\left[-\frac{1}{4}\left(1-x^{\frac{1}{2}}\right)^2 - 2\left(-\frac{1}{2}\right)\left(1-\frac{1}{2}x^{\frac{1}{2}}\right)\left(1-x^{\frac{1}{2}}\right)\right]}{\left(1-x^{\frac{1}{2}}\right)^4} =$$

$$= \frac{-\frac{1}{4}\left(1-x^{\frac{1}{2}}\right)^2 + \left(1-\frac{1}{2}x^{\frac{1}{2}}\right)\left(1-x^{\frac{1}{2}}\right)}{x^{\frac{1}{2}}\cdot\left(1-x^{\frac{1}{2}}\right)^4} = \frac{-\frac{1}{4}\left(1-x^{\frac{1}{2}}\right) + \left(1-\frac{1}{2}x^{\frac{1}{2}}\right)}{x^{\frac{1}{2}}\cdot\left(1-x^{\frac{1}{2}}\right)^3} =$$

$$= \frac{-\frac{1}{4}+\frac{1}{4}x^{\frac{1}{2}}+1-\frac{1}{2}x^{\frac{1}{2}}}{x^{\frac{1}{2}}\cdot\left(1-x^{\frac{1}{2}}\right)^3} = \frac{\frac{3}{4}-\frac{1}{4}x^{\frac{1}{2}}}{x^{\frac{1}{2}}\cdot\left(1-x^{\frac{1}{2}}\right)^3}$$

The second derivative, at the point x where the first derivative is equal to 0, it is:

$$f''(4) = \frac{\frac{3}{4}-\frac{1}{4}\cdot 4^{\frac{1}{2}}}{4^{\frac{1}{2}}\cdot\left(1-4^{\frac{1}{2}}\right)^3} = \frac{\frac{3}{4}-\frac{1}{2}}{2\cdot(1-2)^3} = \frac{\frac{3-2}{4}}{2\cdot(-1)^3} = \frac{\frac{1}{4}}{-2} < 0$$

$x = 4$ is a point of maximum.
The value of the function at the maximum point is:

$$f(4) = \frac{4}{1-\sqrt{4}} = \frac{4}{1-2} = -4$$

11.2.15.5. Inflection Points

We verify whether the function has inflection points, checking whether the second derivative is equal to 0.

$$f''(x) = \frac{\frac{3}{4} - \frac{1}{4}x^{\frac{1}{2}}}{x^{\frac{1}{2}} \cdot \left(1 - x^{\frac{1}{2}}\right)^3} = 0$$

The only solution is that for which the numerator is equal to 0:

$$\frac{3}{4} - \frac{1}{4}x^{\frac{1}{2}} = 0 \quad \Rightarrow \quad \frac{1}{4}x^{\frac{1}{2}} = \frac{3}{4} \quad \Rightarrow \quad x^{\frac{1}{2}} = 3 \quad \Rightarrow \quad x = 9$$

The second derivative is equal to 0 at the point $x = 9$.

The value of the function at the inflection point is:

$$f(9) = \frac{9}{1 - \sqrt{9}} = \frac{9}{1 - 3} = -\frac{9}{2}$$

The graph of the function is shown in Figure 11.15.

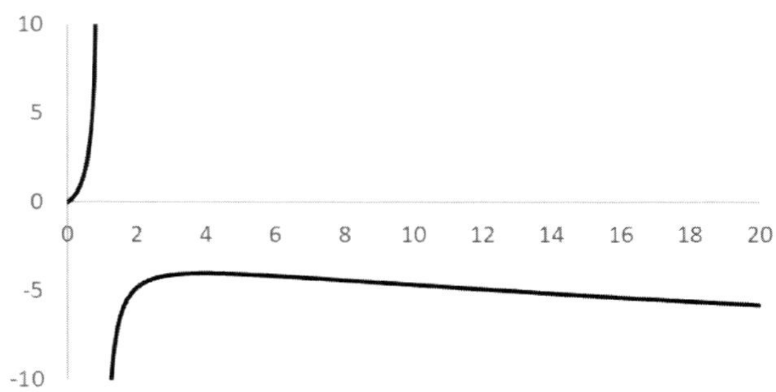

Figure 11.15. Graph of the function $f(x) = \frac{x}{1 - \sqrt{x}}$.

11.2.16. Problem 11.16

Study the function:

$$f(x) = \frac{\log(5-2x)}{5-2x}$$

11.2.16.1. Domain of the Function

The domain of the function is given by the intersection of the function domains in the numerator and the denominator, minus the values in which the denominator is equal to 0.

The logarithm function is defined by:

$$\log : \boldsymbol{R}^{+} \to \boldsymbol{R}$$

In this case, given the argument of the logarithm function we have::

$$5-2x>0 \quad \Rightarrow \quad 5>2x \quad \Rightarrow \quad x<\frac{5}{2}$$

so the domain of the numerator is:

$$D = \left]-\infty, \frac{5}{2}\right[$$

The denominator (its domain is $\boldsymbol{R}$) equals to 0 for:

$$5-2x=0 \quad \Rightarrow \quad x=\frac{5}{2}$$

value already excluded from the domain of the numerator.

The function is therefore defined by:

$$f : \left]-\infty, \frac{5}{2},\right[\to \boldsymbol{R}$$

11.2.16.2. Intersection with the Axes

We verify whether the function intersects the *x*, *y* axes:

$$x=0 \quad \to \quad f(0)=?$$

$$y=f(x)=0 \quad \to \quad x=?$$

For $x = 0$:

$$y = f(0) = \frac{\log(5-2\cdot 0)}{5-2\cdot 0} = \frac{\log 5}{5}$$

The function intersects the y-axis at the point $y = \frac{\log 5}{5}$

For $y = 0$:

$$y = f(x) = \frac{\log(5-2x)}{5-2x} = 0$$

The only solution is that for which the numerator is equal to 0:

$$\log(5-2x) = 0$$

We apply the inverse function of logarithm:

$$e^{\log(5-2x)} = e^0 \quad \Rightarrow \quad 5-2x = 1 \quad \Rightarrow \quad x = \frac{5-1}{2} = 2$$

The function intersects the x-axis at the point $x = 2$, so the function intersects the axes at the points $\left(0, \frac{\log 5}{5}\right)$ and $(2,0)$.

11.2.16.3. Asymptotes

There is a vertical asymptote if:

$$\lim_{x \to x_0+} \frac{\log(5-2x)}{5-2x} = \pm\infty$$

$$\lim_{x \to x_0-} \frac{\log(5-2x)}{5-2x} = \pm\infty$$

Given the function definition domain, we can check the limits:

$$\lim_{x \to \frac{5^-}{2}} \frac{\log(5-2x)}{5-2x} = \frac{\log(5-5^-)}{5-5^-} = \frac{-\infty}{0^+} = -\infty$$

There is a vertical asymptote at $x_0 = \frac{5}{2}$ with one infinite branch to $-\infty$.

Given the function definition domain, for horizontal asymptotes we check the limit only for $x \to -\infty$.

$$\lim_{x\to-\infty} \frac{\log(5-2x)}{5-2x} = l_1,\ l_1 \in \boldsymbol{R}$$

We have:

$$\lim_{x\to-\infty} \frac{\log(5-2x)}{5-2x} = \frac{+\infty}{+\infty}$$

It is an indeterminate form, we can apply De L'Hôpital's rule:

$$\lim_{x\to-\infty} \frac{\log(5-2x)}{5-2x} = \lim_{x\to-\infty} \frac{\frac{d}{dx}\log(5-2x)}{\frac{d}{dx}(5-2x)} =$$

$$= \lim_{x\to-\infty} \frac{\frac{-2}{5-2x}}{-2} = \lim_{x\to-\infty} \frac{1}{5-2x} = \frac{1}{+\infty} = 0$$

For $x \to -\infty$ there is a horizontal asymptote at $y = 0$.

11.2.16.4. Maximum and/or Minimum Points

We verify if the function has maximum or minimum points, checking if the first derivative is equal to 0 at some point and if for these points the second derivative is > 0 or < 0.

$$f'(x_i) = 0$$

$$f''(x_i) > 0 \quad or \quad f''(x_i) < 0$$

First derivative:

$$f'(x) = \frac{d}{dx}\left(\frac{\log(5-2x)}{5-2x}\right) = \frac{\frac{-2}{5-2x}\cdot(5-2x) - \log(5-2x)\cdot(-2)}{(5-2x)^2} =$$
$$= \frac{2\log(5-2x)-2}{(5-2x)^2}$$

$$f'(x) = \frac{2\log(5-2x)-2}{(5-2x)^2} = 0$$

The only solution is that for which the numerator is equal to 0:

$$2\log(5-2x)-2 = 0 \Rightarrow \log(5-2x) = 1 \Rightarrow$$
$$e^{\log(5-2x)} = e^1 \Rightarrow 5-2x = e$$

The first derivative is equal to 0 at the point:

$$x = \frac{5-e}{2}$$

Second derivative:

$$f''(x) = \frac{d}{dx}\left(\frac{2\log(5-2x)-2}{(5-2x)^2}\right) =$$
$$= \frac{2\cdot\frac{-2}{5-2x}\cdot(5-2x)^2 - [2\log(5-2x)-2]\cdot 2(5-2x)\cdot(-2)}{(5-2x)^4} =$$
$$= \frac{-4(5-2x)+4(5-2x)\cdot[2\log(5-2x)-2]}{(5-2x)^4} =$$
$$= \frac{4\cdot[-1+2\log(5-2x)-2]}{(5-2x)^3} = \frac{4\cdot[2\log(5-2x)-3]}{(5-2x)^3}$$

The second derivative, at the point x where the first derivative is equal to 0, it is:

$$f''\left(\frac{5-e}{2}\right)=\frac{4\cdot\left[2\log\left(5-2\cdot\frac{5-e}{2}\right)-3\right]}{\left(5-2\cdot\frac{5-e}{2}\right)^3}=\frac{4\cdot[2\log(5-5+e)-3]}{(5-5+e)^3}=$$

$$=\frac{4\cdot[2\log e-3]}{e^3}=\frac{4\cdot(2-3)}{e^3}<0$$

$x=\frac{5-e}{2}$ is a point of maximum.

The value of the function at the maximum point is:

$$f\left(\frac{5-e}{2}\right)=\frac{\log\left(5-2\cdot\frac{5-e}{2}\right)}{5-2\cdot\frac{5-e}{2}}=\frac{\log e}{e}=e^{-1}$$

11.2.16.5. Inflection Points

We verify whether the function has inflection points, checking whether the second derivative is equal to 0:

$$f''(x)=\frac{4\cdot[2\log(5-2x)-3]}{(5-2x)^3}=0$$

The only solution is that for which the numerator is equal to 0:

$$2\log(5-2x)-3=0\Rightarrow\log(5-2x)=\frac{3}{2}\Rightarrow$$

$$e^{\log(5-2x)}=e^{\frac{3}{2}}\Rightarrow 5-2x=e^{\frac{3}{2}}$$

The second derivative is equal to 0 at the point:

$$x=\frac{5-e^{\frac{3}{2}}}{2}$$

The value of the function at the inflection point is:

$$f\left(\frac{5-e^{\frac{3}{2}}}{2}\right)=\frac{\log\left(5-2\cdot\frac{5-e^{\frac{3}{2}}}{2}\right)}{5-2\cdot\frac{5-e^{\frac{3}{2}}}{2}}=\frac{\log e^{\frac{3}{2}}}{e^{\frac{3}{2}}}=\frac{\frac{3}{2}\log e}{e^{\frac{3}{2}}}=\frac{3}{2}e^{-\frac{3}{2}}$$

The graph of the function is shown in Figure 11.16.

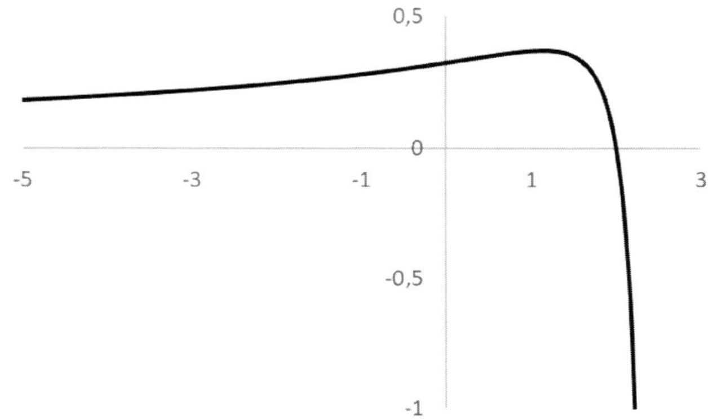

Figure 11.16. Graph of the function $f(x)=\dfrac{\log(5-2x)}{5-2x}$.

11.2.17. Problem 11.17

Study the function:

$$f(x)=\frac{x+1}{\log(x+1)}$$

11.2.17.1. Domain of the Function

The function in the numerator is defined on **R**, so the domain of the function depends on the denominator domain minus the values in which the denominator is equal to 0.

The logarithm function is defined by::

$$\log:\boldsymbol{R}^{+}\rightarrow\boldsymbol{R}$$

In this case, the argument of the logarithm function is:

$$x+1>0 \quad \Rightarrow \quad x>-1$$

so the domain of the denominator is:

$$D=\left]-1,+\infty\right[$$

For the logarithm function it holds $\log 1=0$, so the denominator is equal to 0 for:

$$\log(x+1)=0 \quad \Rightarrow \quad x+1=1 \quad \Rightarrow \quad x=0$$

a value that must be excluded from the domain:

$$D=\left]-1,0\right[\cup\left]0,+\infty\right[$$

The function is therefore defined by:

$$f:\left]-1,0\right[\cup\left]0,+\infty\right[\rightarrow \boldsymbol{R}$$

11.2.17.2. Intersection with the Axes

We verify whether the function intersects the *x*, *y* axes:

$$x=0 \quad \rightarrow \quad f(0)=?$$
$$y=f(x)=0 \quad \rightarrow \quad x=?$$

The point $x=0$ does not belong to the domain of the function, so the function does not intersect the *y*-axis.

For $y=0$:

$$y=f(x)=\frac{x+1}{\log(x+1)}=0$$

The only solution is that for which the numerator is equal to 0:

$$x+1=0 \quad \Rightarrow \quad x=-1$$

The point $x=-1$ does not belong to the domain of the function, so the function does not intersect the *x*-axis.

11.2.17.3. Asymptotes

There is a vertical asymptote if:

$$\lim_{x \to x_0+} \frac{x+1}{\log(x+1)} = \pm\infty$$

$$\lim_{x \to x_0-} \frac{x+1}{\log(x+1)} = \pm\infty$$

Given the domain of the function, let's check the limits for $x_0 = -1^+$, 0^- and 0^+:

$$\lim_{x \to -1+} \frac{x+1}{\log(x+1)} = \frac{0^+}{\log(0^+)} = \frac{0^+}{-\infty} = 0$$

there is no vertical asymptote

$$\lim_{x \to 0-} \frac{x+1}{\log(x+1)} = \frac{1^-}{\log(1^-)} = \frac{1}{0^-} = -\infty$$

$$\lim_{x \to 0+} \frac{x+1}{\log(x+1)} = \frac{1^+}{\log(1^+)} = \frac{1}{0^+} = +\infty$$

There is a vertical asymptote at the point $x_0 = 0$ with two infinite branches, $-\infty$ and $+\infty$.

Given the domain of the function, there is a horizontal asymptote if:

$$\lim_{x \to +\infty} \frac{x+1}{\log(x+1)} = l_1,\ l_1 \in \boldsymbol{R}$$

We have:

$$\lim_{x \to +\infty} \frac{x+1}{\log(x+1)} = \frac{+\infty}{+\infty}$$

It is an indeterminate form, to solve it we can apply De L'Hôpital's rule:

$$\lim_{x\to+\infty}\frac{x+1}{\log(x+1)}=\lim_{x\to+\infty}\frac{\frac{d}{dx}(x+1)}{\frac{d}{dx}\log(x+1)}=\lim_{x\to+\infty}\frac{1}{\frac{1}{x+1}}=\lim_{x\to+\infty}(x+1)=+\infty$$

There is no horizontal asymptote for $x\to+\infty$, so it is necessary to check whether there is an oblique asymptote at the straight line $y = mx + q$. This is true if:

$$\lim_{x\to+\infty}\frac{f(x)}{x}=m \quad \lim_{x\to+\infty}[f(x)-mx]=q \quad m,q\in \boldsymbol{R}$$

For the coefficient m we can proceed simplifying without applying De L'Hôpital's rule.

$$\lim_{x\to+\infty}\frac{f(x)}{x}=\lim_{x\to+\infty}\frac{1}{x}\cdot\frac{x+1}{\log(x+1)}=\lim_{x\to+\infty}\frac{\frac{1}{x}(x+1)}{\log(x+1)}=$$

$$=\lim_{x\to+\infty}\frac{1+\frac{1}{x}}{\log(x+1)}=\frac{1+0}{+\infty}=0$$

Since $m = 0$, the function does not have an oblique asymptote. The result $m = 0$ is sufficient to establish that there is no oblique asymptote. In fact $m = 0$ corresponds to a horizontal straight line, that is a horizontal asymptote that, if there was, it should have been identified previously.

11.2.17.4. Maximum and/or Minimum Points

We verify if the function has maximum or minimum points, checking if the first derivative is equal to 0 at some point and if for these points the second derivative is > 0 or < 0.

$$f'(x_i)=0$$

$$f''(x_i)>0 \quad or \quad f''(x_i)<0$$

First derivative:

$$f'(x)=\frac{d}{dx}\left(\frac{x+1}{\log(x+1)}\right)=\frac{1\cdot\log(x+1)-(x+1)\cdot\dfrac{1}{x+1}}{\log^2(x+1)}=\frac{\log(x+1)-1}{\log^2(x+1)}$$

$$f'(x)=\frac{\log(x+1)-1}{\log^2(x+1)}=0$$

The only solution is that for which the numerator is equal to 0:

$$\log(x+1)-1=0\Rightarrow\log(x+1)=1\Rightarrow$$

$$e^{\log(x+1)}=e^1\Rightarrow x+1=e$$

The first derivative is equal to 0 at the point $x = e - 1$.
Second derivative:

$$f''(x)=\frac{d}{dx}\left(\frac{\log(x+1)-1}{\log^2(x+1)}\right)=$$

$$=\frac{\dfrac{1}{x+1}\cdot\log^2(x+1)-[\log(x+1)-1]\cdot 2\log(x+1)\cdot\dfrac{1}{x+1}}{\log^4(x+1)}=$$

$$=\frac{\dfrac{1}{x+1}\cdot\log(x+1)[\log(x+1)-2\log(x+1)+2]}{\log^4(x+1)}=$$

$$=\frac{[\log(x+1)-2\log(x+1)+2]}{(x+1)\log^3(x+1)}=\frac{2-\log(x+1)}{(x+1)\log^3(x+1)}$$

The second derivative, at the point x where the first derivative is equal to 0, it is:

$$f''(e-1)=\frac{2-\log(e-1+1)}{(e-1+1)\log^3(e-1+1)}=\frac{2-\log e}{e\log^3 e}=\frac{2-1}{e}=e^{-1}>0$$

$x=e-1$ is a point of minimum.
The value of the function at the minimum point is:

$$f(e-1)=\frac{e-1+1}{\log(e-1+1)}=\frac{e}{\log e}=e$$

11.2.17.5. Inflection Points

We verify whether the function has inflection points, checking whether the second derivative is equal to 0.

$$f''(x) = \frac{2 - \log(x+1)}{(x+1)\log^3(x+1)} = 0$$

The only solution is that for which the numerator is equal to 0:

$$2 - \log(x+1) = 0 \Rightarrow \log(x+1) = 2 \Rightarrow$$
$$e^{\log(x+1)} = e^2 \Rightarrow x + 1 = e^2$$

The second derivative is equal to 0 at the point $x = e^2 - 1$.
The value of the function at the inflection point is:

$$f(e^2 - 1) = \frac{e^2 - 1 + 1}{\log(e^2 - 1 + 1)} = \frac{e^2}{\log e^2} = \frac{e^2}{2\log e} = \frac{1}{2}e^2$$

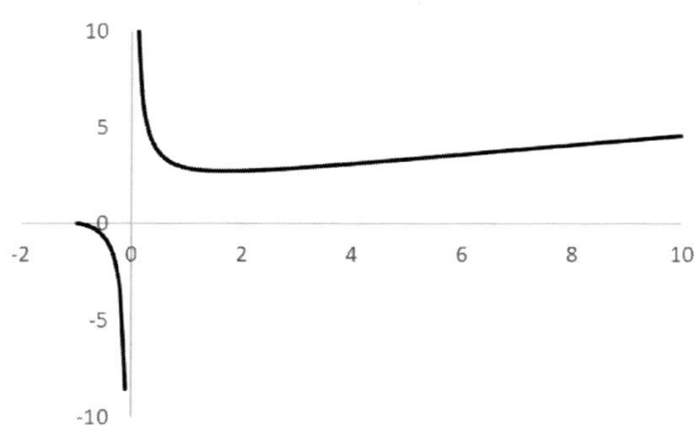

Figure 11.17. Graph of the function $f(x) = \frac{x+1}{\log(x+1)}$.

The graph of the function is shown in Figure 11.17.

11.2.18. Problem 11.18

Study the function:

$$f(x) = \left(e^{-x} - e^{x+1}\right)^2$$

11.2.18.1. Domain of the Function

The function shows the algebraic sum of two exponentials, so the domain is $\boldsymbol{R}$. The function is defined by:

$$f : \boldsymbol{R} \to \boldsymbol{R}^{+}$$

11.2.18.2. Intersection with the Axes

We verify whether the function intersects the *x*, *y* axes:

$$x = 0 \quad \to \quad f(0) = ?$$
$$y = f(x) = 0 \quad \to \quad x = ?$$

For $x = 0$:

$$y = f(0) = \left(e^{-0} - e^{0+1}\right)^2 = (1 - e)^2$$

The function intersects the *y*-axis at the point $y = (1 - e)^2$.

For $y = 0$:

$$y = f(x) = \left(e^{-x} - e^{x+1}\right)^2 = 0$$

The only possible solution is:

$$e^{-x} - e^{x+1} = 0 \Rightarrow e^{-x} = e^{x+1} \Rightarrow \log e^{-x} = \log e^{x+1}$$
$$\Rightarrow -x = x + 1 \Rightarrow 2x = -1 \Rightarrow x = -\frac{1}{2}$$

The function intersects the *x*-axis at the point $x = -\frac{1}{2}$.

11.2.18.3. Asymptotes

There is a vertical asymptote if:

$$\lim_{x \to x_0+} \left(e^{-x} - e^{x+1}\right)^2 = \pm\infty$$

$$\lim_{x \to x_0-} \left(e^{-x} - e^{x+1}\right)^2 = \pm\infty$$

The function is continuous on $\boldsymbol{R}$, so there are no discontinuity points x_0 and the function has no vertical asymptotes.

There is a horizontal asymptote if:

$$\lim_{x \to +\infty} \left(e^{-x} - e^{x+1}\right)^2 = l_1,\ l_1 \in \boldsymbol{R}$$

$$\lim_{x \to -\infty} \left(e^{-x} - e^{x+1}\right)^2 = l_2\ ,\ l_2 \in \boldsymbol{R}$$

We have:

$$\lim_{x \to +\infty} \left(e^{-x} - e^{x+1}\right)^2 = \left(e^{-\infty} - e^{+\infty}\right)^2 = (0 - \infty)^2 = +\infty$$

$$\lim_{x \to -\infty} \left(e^{-x} - e^{x+1}\right)^2 = \left(e^{+\infty} - e^{-\infty}\right)^2 = (+\infty - 0)^2 = +\infty$$

The function has no horizontal asymptotes for $x \to +\infty$ and $x \to -\infty$.

Since there are no horizontal asymptotes, we have to verify if there are oblique asymptotes.

$$\lim_{x \to +\infty} \frac{f(x)}{x} = m_1\ ;\ \lim_{x \to +\infty} \left[f(x) - m_1 x\right] = q_1\ ;\ m_1, q_1 \in \boldsymbol{R}$$

$$\lim_{x \to -\infty} \frac{f(x)}{x} = m_2\ ;\ \lim_{x \to -\infty} \left[f(x) - m_2 x\right] = q_2;\ m_2, q_2 \in \boldsymbol{R}$$

In this case:

$$\lim_{x \to +\infty} \frac{f(x)}{x} = \lim_{x \to +\infty} \frac{\left(e^{-x} - e^{x+1}\right)^2}{x} = \frac{+\infty}{+\infty}$$

It is an indeterminate form. To solve the indeterminate form we can apply De L'Hôpital's rule:

$$\lim_{x\to+\infty}\frac{\left(e^{-x}-e^{x+1}\right)^2}{x}=\lim_{x\to+\infty}\frac{\frac{d}{dx}\left(e^{-x}-e^{x+1}\right)^2}{\frac{d}{dx}(x)}=$$

$$=\lim_{x\to+\infty}\frac{2\left(e^{-x}-e^{x+1}\right)\cdot\left(-e^{-x}-e^{x+1}\right)}{1}=2(-\infty)(-\infty)=+\infty$$

There is no oblique asymptote for $x\to+\infty$.

For $x\to-\infty$ we have:

$$\lim_{x\to-\infty}\frac{f(x)}{x}=\lim_{x\to-\infty}\frac{\left(e^{-x}-e^{x+1}\right)^2}{x}=\frac{+\infty}{-\infty}$$

Also in this case, to resolve the indeterminate form we proceed by applying De L'Hôpital's rule:

$$\lim_{x\to-\infty}\frac{\left(e^{-x}-e^{x+1}\right)^2}{x}=\lim_{x\to-\infty}\frac{\frac{d}{dx}\left(e^{-x}-e^{x+1}\right)^2}{\frac{d}{dx}(x)}=$$

$$=\lim_{x\to-\infty}\frac{2\left(e^{-x}-e^{x+1}\right)\cdot\left(-e^{-x}-e^{x+1}\right)}{1}=2(+\infty)(-\infty)=-\infty$$

There is no oblique asymptote for $x\to-\infty$.

11.2.18.4. Maximum and/or Minimum Points

We verify if the function has maximum or minimum points, checking if the first derivative is equal to 0 at some point and if for these points the second derivative is > 0 or < 0.

$$f'(x_i)=0$$

$$f''(x_i)>0 \quad or \quad f''(x_i)<0$$

First derivative:

$$f'(x) = \frac{d}{dx}\left(e^{-x} - e^{x+1}\right)^2 = 2\left(e^{-x} - e^{x+1}\right)\cdot\left(-e^{-x} - e^{x+1}\right) =$$
$$= -2\left(e^{-x} - e^{x+1}\right)\cdot\left(e^{-x} + e^{x+1}\right) == -2\left[\left(e^{-x}\right)^2 - \left(e^{x+1}\right)^2\right] =$$
$$= -2\left(e^{-2x} - e^{2x+2}\right) = 2\left(e^{2x+2} - e^{-2x}\right)$$
$$f'(x) = 2\left(e^{2x+2} - e^{-2x}\right) = 0$$

The only solution is:

$$e^{2x+2} - e^{-2x} = 0 \Rightarrow e^{2x+2} = e^{-2x} \Rightarrow \log e^{2x+2} = \log e^{-2x} \Rightarrow$$
$$= 2x + 2 = -2x \Rightarrow x + 1 = -x \Rightarrow x = -\frac{1}{2}$$

The first derivative is equal to 0 at the point:

$$x = -\frac{1}{2}$$

Second derivative:

$$f''(x) = \frac{d}{dx} 2\left(e^{2x+2} - e^{-2x}\right) = 2\left(2e^{2x+2} - (-2)e^{-2x}\right) = 4\left(e^{2x+2} + e^{-2x}\right)$$

The second derivative, at the point x where the first derivative is equal to 0, it is:

$$f''\left(-\frac{1}{2}\right) = 4\left(e^{2\left(-\frac{1}{2}\right)+2} + e^{-2\left(-\frac{1}{2}\right)}\right) = 4\left(e^{-1+2} + e^{1}\right) = 4(e + e) = 8e > 0$$

$x = -\frac{1}{2}$ is a point of minimum.

The value of the function at the minimum point is:

$$f\left(-\frac{1}{2}\right) = \left(e^{-\left(-\frac{1}{2}\right)} - e^{-\frac{1}{2}+1}\right)^2 = \left(e^{\frac{1}{2}} - e^{\frac{1}{2}}\right)^2 = 0$$

11.2.18.5. Inflection Points

We verify whether the function has inflection points, checking whether the second derivative is equal to 0:

$$f''(x) = 4\left(e^{2x+2} + e^{-2x}\right) = 0$$

Given the nature of the exponential function and given that only sums of positive values appear in the second derivative, the equation has no solution, so the function has no inflection points.

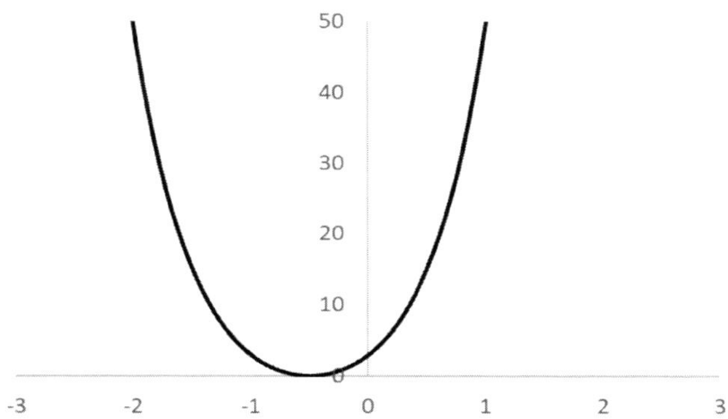

Figure 11.18. Graph of the function $f(x) = \left(e^{-x} - e^{x+1}\right)^2$.

The graph of the function is shown in Figure 11.18.

11.2.19. Problem 11.19

Study the function:

$$f(x) = \frac{1 - \sqrt{x-1}}{x-1}$$

11.2.19.1. Domain of the Function

The domain of the function is given by the intersection of the function domains in the numerator and the denominator, minus the values in which the denominator is equal to 0.

With regard to the square root in the numerator, it is defined by:

$$\sqrt{x-1} \equiv (x-1)^{\frac{1}{2}}: \quad [1,+\infty[\to \boldsymbol{R}$$

The value for which the denominator is equal to 0 it is $x = 1$ which should be excluded from the domain.

The function is therefore defined by:

$$f:]1,+\infty[\to \boldsymbol{R}$$

11.2.19.2. Intersection with the Axes

We verify whether the function intersects the x, y axes:

$$x = 0 \quad \to \quad f(0) = ?$$
$$y = f(x) = 0 \quad \to \quad x = ?$$

For $x = 0$ the function is not defined, so there can be no intersection with the y-axis.
For $y = 0$:

$$y = f(x) = \frac{1-\sqrt{x-1}}{x-1} = 0$$

The only solution is that for which the numerator is equal to 0:

$$1-\sqrt{x-1} = 0 \Rightarrow \sqrt{x-1} = 1 \Rightarrow x-1 = 1 \Rightarrow x = 1+1 = 2$$

The function intersects the x-axis at point $x = 2$.

11.2.19.3. Asymptotes

There is a vertical asymptote if:

$$\lim_{x \to x_0+} \frac{1-\sqrt{x-1}}{x-1} = \pm\infty$$
$$\lim_{x \to x_0-} \frac{1-\sqrt{x-1}}{x-1} = \pm\infty$$

Given the domain of the function, we check the limit only for $x_0^+ = 1^+$:

$$\lim_{x \to 1+} \frac{1-\sqrt{x-1}}{x-1} = \frac{1}{0^+} = +\infty$$

There is a vertical asymptote at the point $x_0 = 1$.

There is a horizontal asymptote if:

$$\lim_{x \to -\infty} \frac{1-\sqrt{x-1}}{x-1} = l_1 \ , \ l_1 \in \boldsymbol{R}$$

$$\lim_{x \to +\infty} \frac{1-\sqrt{x-1}}{x-1} = l_2 \ , \ l_2 \in \boldsymbol{R}$$

Given the domain of the function, we check the limit only for $x \to +\infty$:

$$\lim_{x \to +\infty} \frac{1-\sqrt{x-1}}{x-1} = \frac{-\infty}{+\infty}$$

It is an indeterminate form, to solve it we can apply De L'Hôpital's rule:

$$\lim_{x \to +\infty} \frac{1-\sqrt{x-1}}{x-1} = \lim_{x \to +\infty} \frac{\frac{d}{dx}\left(1-\sqrt{x-1}\right)}{\frac{d}{dx}(x-1)} = \lim_{x \to +\infty} \frac{\frac{d}{dx}\left(1-(x-1)^{\frac{1}{2}}\right)}{\frac{d}{dx}(x-1)} =$$

$$= \lim_{x \to +\infty} \frac{-\frac{1}{2}(x-1)^{-\frac{1}{2}}}{1} = \lim_{x \to +\infty}\left(-\frac{1}{2(x-1)^{\frac{1}{2}}}\right) = -\frac{1}{+\infty} = 0$$

For $x \to +\infty$ there is a horizontal asymptote at $y = 0$.

Since the function is continuous on the definition domain and given the existence of a vertical and a horizontal asymptote, there can be no oblique asymptotes.

11.2.19.4. Maximum and/or Minimum Points

We verify if the function has maximum or minimum points, checking if the first derivative is equal to 0 at some point and if for these points the second derivative is > 0 or < 0.

$$f'(x_i)=0$$
$$f''(x_i)>0 \quad or \quad f''(x_i)<0$$

First derivative:

$$f'(x)=\frac{d}{dx}\left(\frac{1-\sqrt{x-1}}{x-1}\right)\equiv\frac{d}{dx}\left(\frac{1-(x-1)^{\frac{1}{2}}}{x-1}\right)=$$

$$=\frac{-\frac{1}{2}\cdot(x-1)^{-\frac{1}{2}}\cdot(x-1)-\left[1-(x-1)^{\frac{1}{2}}\right]\cdot 1}{(x-1)^2}=$$

$$=\frac{-\frac{1}{2}\cdot(x-1)^{\frac{1}{2}}-1+(x-1)^{\frac{1}{2}}}{(x-1)^2}=\frac{\frac{1}{2}\cdot(x-1)^{\frac{1}{2}}-1}{(x-1)^2}$$

$$f'(x)=\frac{\frac{1}{2}\cdot(x-1)^{\frac{1}{2}}-1}{(x-1)^2}=0$$

The only solution is that for which the numerator is equal to 0:

$$\frac{1}{2}\cdot(x-1)^{\frac{1}{2}}-1=0\Rightarrow\frac{1}{2}\cdot(x-1)^{\frac{1}{2}}=1\Rightarrow(x-1)^{\frac{1}{2}}=2\Rightarrow$$
$$x-1=4\Rightarrow x=5$$

The first derivative is equal to 0 at the point $x = 5$.
Second derivative:

$$f''(x) = \frac{d}{dx}\left(\frac{\frac{1}{2}\cdot(x-1)^{\frac{1}{2}} - 1}{(x-1)^2}\right) =$$

$$= \frac{\frac{1}{2}\cdot\frac{1}{2}\cdot(x-1)^{-\frac{1}{2}}\cdot(x-1)^2 - \left(\frac{1}{2}\cdot(x-1)^{\frac{1}{2}} - 1\right)\cdot 2(x-1)}{(x-1)^4} =$$

$$= \frac{(x-1)\cdot\left[\frac{1}{4}(x-1)^{-\frac{1}{2}}\cdot(x-1) - \left(\frac{1}{2}\cdot(x-1)^{\frac{1}{2}} - 1\right)\cdot 2\right]}{(x-1)^4} =$$

$$= \frac{\frac{1}{4}(x-1)^{1-\frac{1}{2}} - \frac{1}{2}\cdot 2(x-1)^{\frac{1}{2}} + 2}{(x-1)^3} =$$

$$= \frac{\frac{1}{4}(x-1)^{\frac{1}{2}} - (x-1)^{\frac{1}{2}} + 2}{(x-1)^3} = \frac{-\frac{3}{4}(x-1)^{\frac{1}{2}} + 2}{(x-1)^3}$$

The second derivative, at the point x where the first derivative is equal to 0, it is:

$$f''(5) = \frac{-\frac{3}{4}(5-1)^{\frac{1}{2}} + 2}{(5-1)^3} = \frac{-\frac{3}{4}\cdot 2 + 2}{4^3} = \frac{-\frac{3}{2} + 2}{64} = \frac{\frac{1}{2}}{64} > 0$$

$x = 5$ is a point of minimum.

The value of the function at the minimum point is:

$$f(5) = \frac{1 - \sqrt{5-1}}{5-1} = \frac{1-2}{4} = -\frac{1}{4}$$

11.2.19.5. Inflection Points

We verify whether the function has inflection points, checking whether the second derivative is equal to 0.

$$f''(x) = \frac{-\frac{3}{4}(x-1)^{\frac{1}{2}} + 2}{(x-1)^3} = 0$$

The only solution is that for which the numerator is equal to 0:

$$-\frac{3}{4}(x-1)^{\frac{1}{2}}+2=0\Rightarrow\frac{3}{4}(x-1)^{\frac{1}{2}}=2\Rightarrow(x-1)^{\frac{1}{2}}=2\cdot\frac{4}{3}=\frac{8}{3}\Rightarrow$$

$$x-1=\left(\frac{8}{3}\right)^2=\frac{64}{9}\Rightarrow x=\frac{64}{9}+1=\frac{73}{9}$$

The second derivative is equal to 0 at the point $x=\frac{73}{9}$.

The value of the function at the inflection point is:

$$f\left(\frac{73}{9}\right)=\frac{1-\sqrt{\frac{73}{9}-1}}{\frac{73}{9}-1}=\frac{1-\sqrt{\frac{64}{9}}}{\frac{64}{9}}=\frac{1-\frac{8}{3}}{\frac{64}{9}}=\frac{-\frac{5}{3}}{\frac{64}{9}}=-\frac{5}{3}\cdot\frac{9}{64}=-\frac{15}{64}$$

The graph of the function is shown in Figure 11.19.

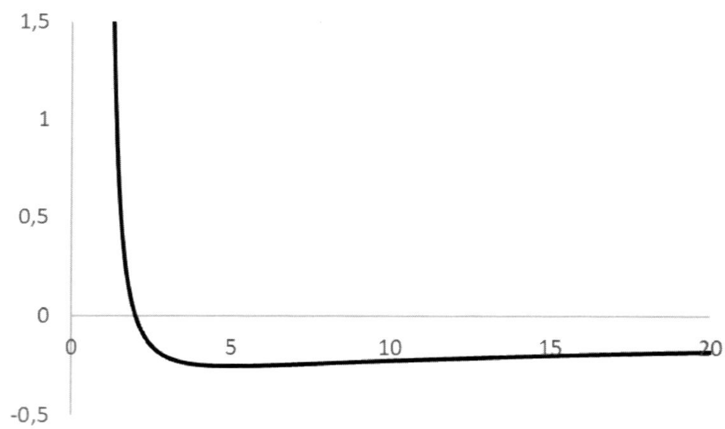

Figure 11.19. Graph of the function $f(x)=\frac{1-\sqrt{x-1}}{x-1}$.

11.2.20. Problem 11.20

Study the function:

$$f(x)=\frac{x}{(x-2)^2}$$

11.2.20.1. Domain of the Function

The domain of the function is given by the intersection of the function domains in the numerator and the denominator, in this case $\boldsymbol{R}$, minus the values in which the denominator is equal to 0.

The domain of the function is therefore determined by the values for which the denominator is not equal to 0:

$$x-2\neq 0 \quad \Rightarrow \quad x\neq 2$$

The function is therefore defined by:

$$f:\boldsymbol{R}-\{2\}\rightarrow \boldsymbol{R}$$

11.2.20.2. Intersection with the Axes

We verify whether the function intersects the x, y axes:

$$x=0 \quad \rightarrow \quad f(0)=?$$
$$y=f(x)=0 \quad \rightarrow \quad x=?$$

For $x=0$:

$$y=f(0)=\frac{0}{(0-2)^2}=0$$

The function intersects the y-axis at the point $y=0$.
For $y=0$:

$$y=f(x)=\frac{x}{(x-2)^2}=0$$

The only solution is that for which the numerator is equal to 0:

$$x=0$$

The function intersects the x-axis at the point $x=0$, so the function intersects the axes at the origin (0,0).

11.2.20.3. Asymptotes

There is a vertical asymptote if:

$$\lim_{x \to x_0+} \frac{x}{(x-2)^2} = \pm\infty$$

$$\lim_{x \to x_0-} \frac{x}{(x-2)^2} = \pm\infty$$

Given the domain of the function, let's check the limits for $x_0 = 2$:

$$\lim_{x \to 2+} \frac{x}{(x-2)^2} = \frac{2}{0^+} = +\infty$$

$$\lim_{x \to 2-} \frac{x}{(x-2)^2} = \frac{2}{0^+} = +\infty$$

Where 0^+ is determined by the square of $(x - 2)$.

There is a vertical asymptote at $x = 2$ with two infinite branches to $+\infty$.

There is a horizontal asymptote if:

$$\lim_{x \to -\infty} \frac{x}{(x-2)^2} = l_1 \text{ , } l_1 \in \boldsymbol{R}$$

$$\lim_{x \to +\infty} \frac{x}{(x-2)^2} = l_2 \text{ , } l_2 \in \boldsymbol{R}$$

We have:

$$\lim_{x \to +\infty} \frac{x}{(x-2)^2} = \frac{+\infty}{+\infty}$$

It is an indeterminate form. There is no only method to resolve the indeterminate form. For example, we can proceed as follows:

$$\lim_{x\to+\infty}\frac{x}{(x-2)^2}=\lim_{x\to+\infty}\frac{x}{x^2\left(1-\frac{2}{x}\right)^2}=\lim_{x\to+\infty}\frac{1}{x\left(1-\frac{2}{x}\right)^2}=$$

$$=\frac{1}{+\infty\left(1-\frac{2}{+\infty}\right)^2}=\frac{1}{+\infty}=0$$

Alternatively, the limit can be calculated by applying De L'Hôpital's rule:

$$\lim_{x\to+\infty}\frac{x}{(x-2)^2}=\lim_{x\to+\infty}\frac{\frac{d}{dx}x}{\frac{d}{dx}(x-2)^2}=\lim_{x\to+\infty}\frac{1}{2(x-2)}=\frac{1}{+\infty}=0$$

For $x\to+\infty$ there is a horizontal asymptote at $y=0$.
For $x\to-\infty$ we have:

$$\lim_{x\to-\infty}\frac{x}{(x-2)^2}=\frac{-\infty}{+\infty}$$

It is an indeterminate form, we can apply De L'Hôpital's rule:

$$\lim_{x\to-\infty}\frac{x}{(x-2)^2}=\lim_{x\to-\infty}\frac{\frac{d}{dx}x}{\frac{d}{dx}(x-2)^2}=\lim_{x\to-\infty}\frac{1}{2(x-2)}=\frac{1}{-\infty}=0$$

For $x\to-\infty$ there is a horizontal asymptote at $y=0$.

11.2.20.4. Maximum and/or Minimum Points

We verify if the function has maximum or minimum points, checking if the first derivative is equal to 0 at some point and if for these points the second derivative is > 0 or < 0.

$$f'(x_i)=0$$
$$f''(x_i)>0 \quad or \quad f''(x_i)<0$$

First derivative:

$$f'(x)=\frac{d}{dx}\left(\frac{x}{(x-2)^2}\right)=\frac{1\cdot(x-2)^2-x\cdot 2(x-2)}{(x-2)^4}=$$

$$=\frac{(x-2)\cdot[(x-2)-2x]}{(x-2)^4}=\frac{x-2-2x}{(x-2)^3}=\frac{-x-2}{(x-2)^3}=-\frac{x+2}{(x-2)^3}$$

$$f'(x)=-\frac{x+2}{(x-2)^3}=0$$

The only solution is that for which the numerator is equal to 0:

$$x+2=0 \quad \Rightarrow \quad x=-2$$

The first derivative is equal to 0 at the point $x = -2$.
Second derivative:

$$f''(x)=\frac{d}{dx}\left(-\frac{x+2}{(x-2)^3}\right)=\frac{-1\cdot(x-2)^3-[-(x+2)]\cdot 3(x-2)^2}{(x-2)^6}=$$

$$=\frac{(x-2)^2\cdot[-(x-2)+3(x+2)]}{(x-2)^6}=\frac{-(x-2)+3(x+2)}{(x-2)^4}=$$

$$=\frac{-x+2+3x+6}{(x-2)^4}=\frac{2x+8}{(x-2)^4}=\frac{2(x+4)}{(x-2)^4}$$

The second derivative, at the point x where the first derivative is equal to 0, it is:

$$f''(-2)=\frac{2(-2+4)}{(-2-2)^4}=\frac{4}{(-4)^4}=\frac{4}{256}=\frac{1}{64}>0$$

$x=-2$ is a point of minimum.
The value of the function at the minimum point is:

$$f(-2)=\frac{-2}{(-2-2)^2}=\frac{-2}{4^2}=\frac{-2}{16}=-\frac{1}{8}$$

11.2.20.5. Inflection Points

We verify whether the function has inflection points, checking whether the second derivative is equal to 0.

$$f''(x)=\frac{2(x+4)}{(x-2)^4}=0$$

The only solution is:

$$x+4=0 \quad \Rightarrow \quad x=-4$$

$x = -4$ is an inflection point.
The value of the function at the inflection point is:

$$f(-4)=\frac{-4}{(-4-2)^2}=\frac{-4}{(-6)^2}=\frac{-4}{36}=-\frac{1}{9}$$

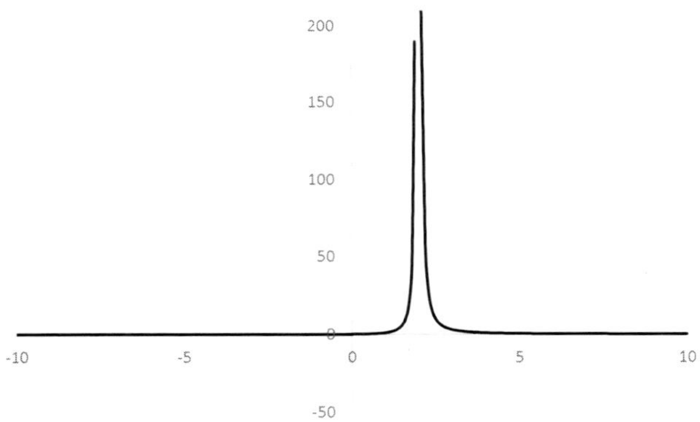

Figure 11.20a. Graph of the function $f(x)=\dfrac{x}{(x-2)^2}$.

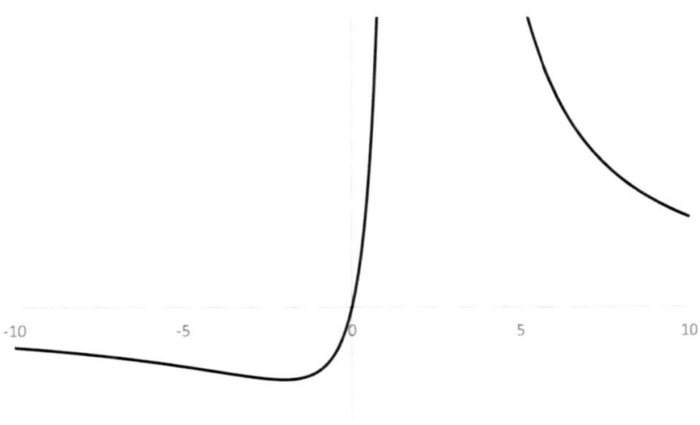

Figure 11.20b. Graph of the function $f(x)=\dfrac{x}{(x-2)^2}$, with a scale suitable for highlighting the minimum and inflection points.

The graph of the function, with a choice of the axes scale such as to show the asymptotes and the points of interest, is shown in Figures 11.20a and 11.20b.

11.1.21. Problem 11.21

Study the function:

$$f(x) = \frac{x+1}{(x-1)^2}$$

11.2.21.1. Domain of the Function

The domain of the function is given by the intersection of the function domains in the numerator and the denominator, in this case ***R***, minus the values in which the denominator is equal to 0.

The domain of the function is therefore determined by the values for which the denominator is not equal to 0:

$$x - 1 \neq 0 \quad \Rightarrow \quad x \neq 1$$

The function is therefore defined by:

$$f : \boldsymbol{R} - \{1\} \to \boldsymbol{R}$$

11.2.21.2. Intersection with the Axes

We verify whether the function intersects the x, y axes:

$$x = 0 \quad \to \quad f(0) = ?$$
$$y = f(x) = 0 \quad \to \quad x = ?$$

For $x = 0$:

$$y = f(0) = \frac{0+1}{(0-1)^2} = 1$$

The function intersects the y-axis at the point $y = 1$.

For $y = 0$:

$$y = f(x) = \frac{x+1}{(x-1)^2} = 0$$

The only solution is that for which the numerator is equal to 0:

$$x+1=0 \quad \Rightarrow \quad x=-1$$

The function intersects the x-axis at the point $x = -1$, so the function intersects the axes at the points $(0,1)$ and $(-1,0)$.

11.2.21.3. Asymptotes

There is a vertical asymptote if:

$$\lim_{x \to x_0+} \frac{x+1}{(x-1)^2} = \pm\infty$$

$$\lim_{x \to x_0-} \frac{x+1}{(x-1)^2} = \pm\infty$$

Given the domain of the function, let's check the limits for $x_0 = 1$:

$$\lim_{x \to 1+} \frac{x+1}{(x-1)^2} = \frac{2}{0^+} = +\infty$$

$$\lim_{x \to 1-} \frac{x+1}{(x-1)^2} = \frac{2}{0^+} = +\infty$$

Where 0^+ is determined by the square of $(x - 1)$.

There is a vertical asymptote at $x = 1$ with two infinite branches to $+\infty$.

There is a horizontal asymptote if:

$$\lim_{x \to -\infty} \frac{x+1}{(x-1)^2} = l_1 \ , \ l_1 \in \boldsymbol{R}$$

$$\lim_{x \to +\infty} \frac{x+1}{(x-1)^2} = l_2 \ , \ l_2 \in \boldsymbol{R}$$

We have:

$$\lim_{x\to+\infty}\frac{x+1}{(x-1)^2}=\frac{+\infty}{+\infty}$$

It is an indeterminate form. To resolve the indeterminate form we can apply De L'Hôpital's rule.

$$\lim_{x\to+\infty}\frac{x+1}{(x-1)^2}=\lim_{x\to+\infty}\frac{\frac{d}{dx}(x+1)}{\frac{d}{dx}(x-1)^2}=\lim_{x\to+\infty}\frac{1}{2(x-1)}=\frac{1}{+\infty}=0$$

For $x\to+\infty$ there is a horizontal asymptote at $y=0$.

For $x\to-\infty$ we have:

$$\lim_{x\to-\infty}\frac{x+1}{(x-1)^2}=\frac{-\infty}{+\infty}$$

It is an indeterminate form, we can apply De L'Hôpital's rule:

$$\lim_{x\to-\infty}\frac{x+1}{(x-1)^2}=\lim_{x\to-\infty}\frac{\frac{d}{dx}(x+1)}{\frac{d}{dx}(x-1)^2}=\lim_{x\to-\infty}\frac{1}{2(x-1)}=\frac{1}{-\infty}=0$$

For $x\to-\infty$ there is a horizontal asymptote at $y=0$.

11.2.21.4. Maximum and/or Minimum Points

We verify if the function has maximum or minimum points, checking if the first derivative is equal to 0 at some point and if for these points the second derivative is > 0 or < 0.

$$f'(x_i)=0$$

$$f''(x_i)>0 \quad \textit{or} \quad f''(x_i)<0$$

First derivative:

$$f'(x)=\frac{d}{dx}\left(\frac{x+1}{(x-1)^2}\right)=\frac{1\cdot(x-1)^2-(x+1)\cdot 2(x-1)}{(x-1)^4}=$$

$$=\frac{(x-1)\cdot[(x-1)-2(x+1)]}{(x-1)^4}=\frac{x-1-2x-2}{(x-1)^3}=\frac{-x-3}{(x-1)^3}=-\frac{x+3}{(x-1)^3}$$

$$f'(x)=-\frac{x+3}{(x-1)^3}=0$$

The only solution is that for which the numerator is equal to 0:

$$x+3=0 \quad \Rightarrow \quad x=-3$$

The first derivative is equal to 0 at the point $x=-3$.
Second derivative:

$$f''(x)=\frac{d}{dx}\left(-\frac{x+3}{(x-1)^3}\right)=-\frac{1\cdot(x-1)^3-(x+3)\cdot 3(x-1)^2}{(x-1)^6}=$$

$$=-\frac{(x-1)^2\cdot[(x-1)-3(x+3)]}{(x-1)^6}=-\frac{x-1-3x-9}{(x-1)^4}=$$

$$=-\frac{-2x-10}{(x-1)^4}=\frac{2(x+5)}{(x-1)^4}$$

The second derivative, at the point x where the first derivative is equal to 0, it is:

$$f''(-3)=\frac{2\cdot(-3+5)}{(-3-1)^4}=\frac{2\cdot 2}{(-4)^4}=\frac{4}{256}=\frac{1}{64}>0$$

$x=-3$ is a point of minimum.
The value of the function at the minimum point is:

$$f(-3)=\frac{-3+1}{(-3-1)^2}=\frac{-2}{(-4)^2}=-\frac{2}{16}=-\frac{1}{8}$$

11.2.21.5. Inflection Points

We verify whether the function has inflection points, checking whether the second derivative is equal to 0.

$$f''(x) = \frac{2(x+5)}{(x-1)^4} = 0$$

The only solution is:

$$x + 5 = 0 \quad \Rightarrow \quad x = -5$$

$x = -5$ is an inflection point.
The value of the function at the inflection point is:

$$f(-5) = \frac{-5+1}{(-5-1)^2} = \frac{-4}{(-6)^2} = -\frac{4}{36} = -\frac{1}{9}$$

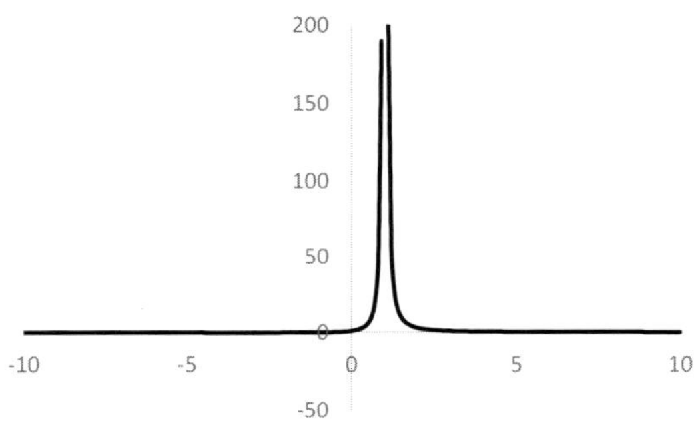

Figure 11.21a. Graph of the function $f(x) = \dfrac{x+1}{(x-1)^2}$.

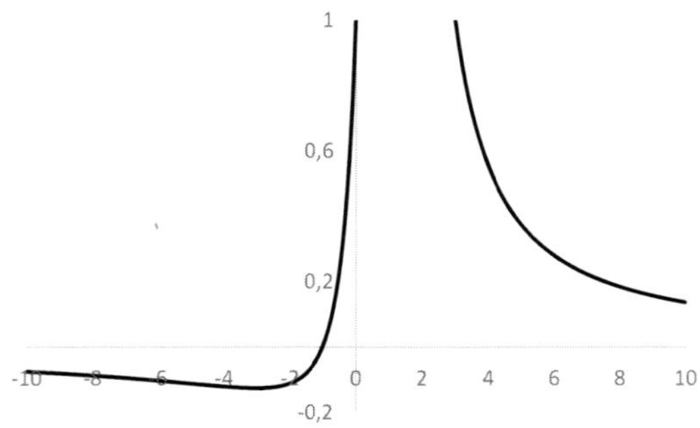

Figure 11.21b. Graph of the function $f(x) = \dfrac{x+1}{(x-1)^2}$, with a scale suitable for highlighting the minimum and inflection points.

The graph of the function, with a choice of the axes scale such as to show the asymptotes and the points of interest, is shown in Figures 11.21a and 11.21b.

11.2.22. Problem 11.22

Study the function:

$$f(x)=\frac{1-x}{(x-3)^2}$$

11.2.22.1. Domain of the Function

The domain of the function is given by the intersection of the function domains in the numerator and the denominator, in this case ***R***, minus the values in which the denominator is equal to 0.

The domain of the function is therefore determined by the values for which the denominator is not equal to 0:

$$x-3\neq 0 \quad \rightarrow \quad x\neq 3$$

The function is therefore defined by:

$$f:\boldsymbol{R}-\{3\}\rightarrow \boldsymbol{R}$$

11.2.22.2. Intersection with the Axes

We verify whether the function intersects the x, y axes:

$$x=0 \quad \rightarrow \quad f(0)=?$$
$$y=f(x)=0 \quad \rightarrow \quad x=?$$

For $x = 0$:

$$y=f(0)=\frac{1-0}{(0-3)^2}=\frac{1}{9}$$

The function intersects the y-axis at the point $y=\frac{1}{9}$.

For $y = 0$:

$$y=f(x)=\frac{1-x}{(x-3)^2}=0$$

The only solution is that for which the numerator is equal to 0:

$$1 - x = 0 \quad \Rightarrow \quad x = 1$$

The function intersects the x-axis at the point $x = 1$, so the function intersects the axes at the points $\left(0, \frac{1}{9}\right)$ and $(1,0)$.

11.2.22.3 Asymptotes

There is a vertical asymptote if.

$$\lim_{x \to x_0 +} \frac{1-x}{(x-3)^2} = \pm\infty$$

$$\lim_{x \to x_0 -} \frac{1-x}{(x-3)^2} = \pm\infty$$

Given the domain of the function, let's check the limits for $x_0 = 3$:

$$\lim_{x \to 3+} \frac{1-x}{(x-3)^2} = \frac{-2}{0^+} = -\infty$$

$$\lim_{x \to 3-} \frac{1-x}{(x-3)^2} = \frac{-2}{0^+} = -\infty$$

Where 0^+ is determined by the square of $(x - 3)$.
There is a vertical asymptote at $x = 3$ with two infinite branches to $-\infty$.
There is a horizontal asymptote if:

$$\lim_{x \to -\infty} \frac{1-x}{(x-3)^2} = l_1 \ , \ l_1 \in \boldsymbol{R}$$

$$\lim_{x \to +\infty} \frac{1-x}{(x-3)^2} = l_2 \ , \ l_2 \in \boldsymbol{R}$$

We have:

$$\lim_{x \to +\infty} \frac{1-x}{(x-3)^2} = \frac{-\infty}{+\infty}$$

It is an indeterminate form. To resolve the indeterminate form we can apply De L'Hôpital's rule.

$$\lim_{x\to+\infty}\frac{1-x}{(x-3)^2}=\lim_{x\to+\infty}\frac{\frac{d}{dx}(1-x)}{\frac{d}{dx}(x-3)^2}=\lim_{x\to+\infty}\frac{-1}{2(x-3)}=\frac{-1}{+\infty}=0$$

For $x\to+\infty$ there is a horizontal asymptote at $y=0$.

For $x\to-\infty$ we have:

$$\lim_{x\to-\infty}\frac{1-x}{(x-3)^2}=\frac{+\infty}{+\infty}$$

It is an indeterminate form, we can apply De L'Hôpital's rule:

$$\lim_{x\to-\infty}\frac{1-x}{(x-3)^2}=\lim_{x\to-\infty}\frac{\frac{d}{dx}(1-x)}{\frac{d}{dx}(x-3)^2}=\lim_{x\to-\infty}\frac{-1}{2(x-3)}=\frac{-1}{-\infty}=0$$

For $x\to-\infty$ there is a horizontal asymptote at $y=0$

11.2.22.4. Maximum and/or Minimum Points

We verify if the function has maximum or minimum points, checking if the first derivative is equal to 0 at some point and if for these points the second derivative is > 0 or < 0.

$$f'(x_i)=0$$

$$f''(x_i)>0 \quad or \quad f''(x_i)<0$$

First derivative:

$$f'(x) = \frac{d}{dx}\left(\frac{1-x}{(x-3)^2}\right) = \frac{-1\cdot(x-3)^2-(1-x)\cdot 2(x-3)}{(x-3)^4} =$$

$$= \frac{(x-3)\cdot[-(x-3)-2(1-x)]}{(x-3)^4} = \frac{-x+3-2+2x}{(x-3)^3} = \frac{x+1}{(x-3)^3}$$

$$f'(x) = \frac{x+1}{(x-3)^3} = 0$$

The only solution is that for which the numerator is equal to 0:

$$x+1=0 \quad \Rightarrow \quad x=-1$$

The first derivative is equal to 0 at the point $x = -1$.
Second derivative:

$$f''(x) = \frac{d}{dx}\left(\frac{x+1}{(x-3)^3}\right) = \frac{1\cdot(x-3)^3-(x+1)\cdot 3(x-3)^2}{(x-3)^6} =$$

$$= \frac{(x-3)^2\cdot[(x-3)-3(x+1)]}{(x-3)^6} = \frac{(x-3)-3(x+1)}{(x-3)^4} = \frac{x-3-3x-3}{(x-3)^4} =$$

$$= \frac{-2x-6}{(x-3)^4} = -\frac{2(x+3)}{(x-3)^4}$$

The second derivative, at the point x where the first derivative is equal to 0, it is:

$$f''(-1) = -\frac{2\cdot(-1+3)}{(-1-3)^4} = -\frac{4}{(-4)^4} = -\frac{4}{256} = -\frac{1}{64} < 0$$

$x = -1$ is a point of maximum.
The value of the function at the maximum point is:

$$f(-1) = \frac{1-(-1)}{(-1-3)^2} = \frac{2}{(-4)^2} = \frac{2}{16} = \frac{1}{8}$$

11.2.22.5. Inflection Points

We verify whether the function has inflection points, checking whether the second derivative is equal to 0.

$$f''(x) = -\frac{2(x+3)}{(x-3)^4} = 0$$

The only solution is:

$$x+3=0 \quad \Rightarrow \quad x=-3$$

$x = -3$ is an inflection point.
The value of the function at the inflection point is::

$$f(-3) = \frac{1-(-3)}{(-3-3)^2} = \frac{4}{(-6)^2} = \frac{4}{36} = \frac{1}{9}$$

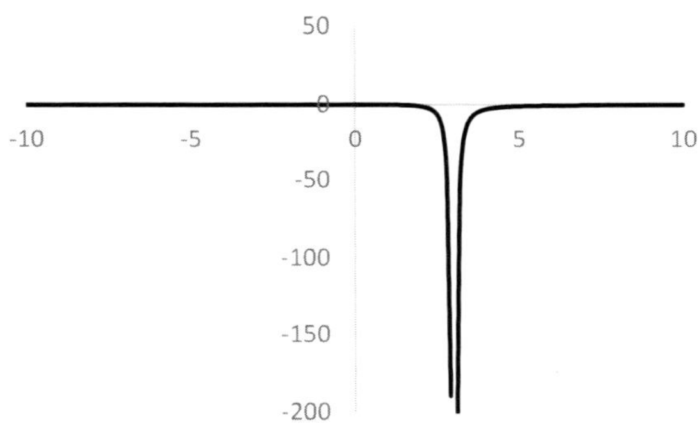

Figure 11.22a. Graph of the function $f(x) = \frac{1-x}{(x-3)^2}$.

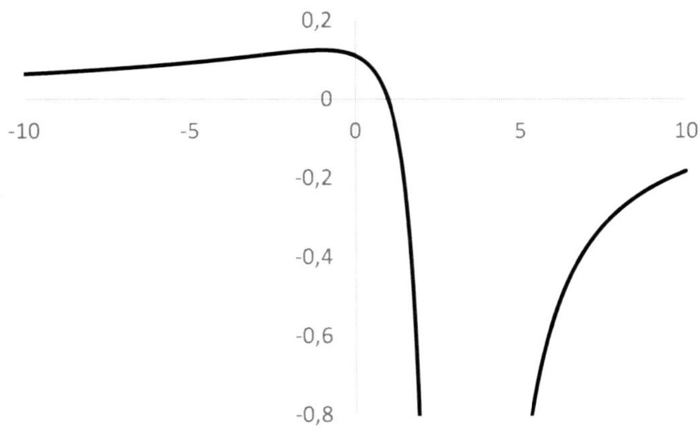

Figure 11.22b. Graph of the function $f(x) = \frac{1-x}{(x-3)^2}$, with a scale suitable for highlighting the minimum and inflection points.

The graph of the function, with a choice of the axes scale such as to show the asymptotes and the points of interest, is shown in Figures 11.22a and 11.22b.

11.2.23. Problem 11.23

Study the function:

$$f(x) = \frac{e^x}{(1-x)^2}$$

11.2.23.1. Domain of the Function

The domain of the function is given by the intersection of the function domains in the numerator and the denominator, in this case ***R***, minus the values in which the denominator is equal to 0.

The domain of the function is therefore determined by the values for which the denominator is not equal to 0:

$$1 - x \neq 0 \quad \Rightarrow \quad x \neq 1$$

The function is therefore defined by:

$$f : \boldsymbol{R} - \{1\} \to \boldsymbol{R}$$

11.2.23.2. Intersection with the Axes

We verify whether the function intersects the x, y axes:

$$x = 0 \quad \rightarrow \quad f(0) = ?$$

$$y = f(x) = 0 \quad \rightarrow \quad x = ?$$

For $x = 0$:

$$y = f(0) = \frac{e^0}{(1-0)^2} = 1$$

The function intersects the y-axis at the point $y = 1$.

For $y = 0$:

$$y = f(x) = \frac{e^x}{(1-x)^2} = 0$$

The equation does not admit a solution, so the function does not intersect the x-axis.

11.2.23.3. Asymptotes

There is a vertical asymptote if:

$$\lim_{x \to x_0 +} \frac{e^x}{(1-x)^2} = \pm\infty$$

$$\lim_{x \to x_0 -} \frac{e^x}{(1-x)^2} = \pm\infty$$

Given the domain of the function, let's check the limits for $x_0 = 1$:

$$\lim_{x \to 1-} \frac{e^x}{(1-x)^2} = \frac{e}{0^+} = +\infty$$

$$\lim_{x \to 1+} \frac{e^x}{(1-x)^2} = \frac{e}{0^+} = +\infty$$

Where 0^+ is determined by the square of $(1 - x)$.
There is a vertical asymptote at $x = 1$ with two infinite branches to $+\infty$.
There is a horizontal asymptote if:

$$\lim_{x \to -\infty} \frac{e^x}{(1-x)^2} = l_1 \ , \ l_1 \in \boldsymbol{R}$$

$$\lim_{x \to +\infty} \frac{e^x}{(1-x)^2} = l_2 \ , \ l_2 \in \boldsymbol{R}$$

We have:

$$\lim_{x \to -\infty} \frac{e^x}{(1-x)^2} = \frac{e^{-\infty}}{(1+\infty)^2} = \frac{0}{+\infty} = 0$$

For $x \to -\infty$ there is a horizontal asymptote at $y = 0$.

For $x \to +\infty$ we have:

$$\lim_{x\to+\infty} \frac{e^x}{(1-x)^2} = \frac{e^{+\infty}}{(1-\infty)^2} = \frac{+\infty}{+\infty}$$

It is an indeterminate form that can be solved and calculated by applying De L'Hôpital's rule:

$$\lim_{x\to+\infty} \frac{e^x}{(1-x)^2} = \lim_{x\to+\infty} \frac{\frac{d}{dx}\left(e^x\right)}{\frac{d}{dx}(1-x)^2} = \lim_{x\to+\infty} \frac{e^x}{-2(1-x)} = \frac{+\infty}{+\infty}$$

After the first derivation the limit still has an indeterminate form, so we can continue to apply De L'Hôpital's rule:

$$\lim_{x\to+\infty} \frac{e^x}{-2(1-x)} = \lim_{x\to+\infty} \frac{\frac{d}{dx}\left(e^x\right)}{\frac{d}{dx}[-2(1-x)]} = \lim_{x\to+\infty} \frac{e^x}{2} = +\infty$$

There is no horizontal asymptote for $x \to +\infty$, so it is necessary to check whether there is an oblique asymptote:

$$\lim_{x\to+\infty} \frac{f(x)}{x} = \lim_{x\to+\infty} \frac{\frac{e^x}{(1-x)^2}}{x} = \lim_{x\to+\infty} \frac{e^x}{x(1-x)^2} = m,\ m \in \mathbf{R}$$

We have:

$$\lim_{x\to+\infty} \frac{e^x}{x(1-x)^2} = \frac{e^{+\infty}}{+\infty \cdot (1-\infty)^2} = \frac{+\infty}{+\infty}$$

It is an indeterminate form, we can apply De L'Hôpital's rule:

$$\lim_{x\to+\infty}\frac{e^x}{x(1-x)^2}=\lim_{x\to+\infty}\frac{\frac{d}{dx}\left(e^x\right)}{\frac{d}{dx}\left[x(1-x)^2\right]}=$$

$$=\lim_{x\to+\infty}\frac{e^x}{(1-x)^2-2x(1-x)}=\frac{+\infty}{+\infty-\infty+\infty}$$

It is still an indeterminate form, so we can continue to apply De L'Hôpital's rule:

$$\lim_{x\to+\infty}\frac{e^x}{(1-x)^2-2x(1-x)}=\lim_{x\to+\infty}\frac{\frac{d}{dx}\left(e^x\right)}{\frac{d}{dx}\left[(1-x)^2-2x(1-x)\right]}=$$

$$=\lim_{x\to+\infty}\frac{e^x}{-2(1-x)-2(1-x)+2x}=\lim_{x\to+\infty}\frac{e^x}{-2+2x-2+2x+2x}=$$

$$=\lim_{x\to+\infty}\frac{e^x}{6x-4}=\frac{+\infty}{+\infty}$$

By continuing to apply De L'Hôpital's rule we have:

$$\lim_{x\to+\infty}\frac{e^x}{6x-4}=\lim_{x\to+\infty}\frac{\frac{d}{dx}\left(e^x\right)}{\frac{d}{dx}(6x-4)}=\lim_{x\to+\infty}\frac{e^x}{6}=+\infty$$

There is no oblique asymptote for $x \to +\infty$.

11.2.23.4. Maximum and/or Minimum Points

We verify if the function has maximum or minimum points, checking if the first derivative is equal to 0 at some point and if for these points the second derivative is > 0 or < 0.

$$f'(x_i)=0$$

$$f''(x_i)>0 \quad or \quad f''(x_i)<0$$

First derivative:

$$f'(x) = \frac{d}{dx}\left(\frac{e^x}{(1-x)^2}\right) = \frac{e^x \cdot (1-x)^2 - e^x(-2)(1-x)}{(1-x)^4} =$$

$$= \frac{e^x(1-x)\cdot(1-x+2)}{(1-x)^4} = \frac{e^x(3-x)}{(1-x)^3}$$

$$f'(x) = \frac{e^x(3-x)}{(1-x)^3} = 0$$

The only solution is that for which the numerator is equal to 0:

$$3 - x = 0 \quad \Rightarrow \quad x = 3$$

The first derivative is equal to 0 at the point $x = 3$.
Second derivative:

$$f''(x) = \frac{d}{dx}\left(\frac{e^x(3-x)}{(1-x)^3}\right) =$$

$$\frac{[e^x(3-x) - e^x](1-x)^3 - e^x(3-x)(-3)(1-x)^2}{(1-x)^6} =$$

$$= \frac{(1-x)^2[(3e^x - xe^x - e^x)(1-x) + 3e^x(3-x)]}{(1-x)^6} =$$

$$= \frac{[(2e^x - xe^x)(1-x) + 3e^x(3-x)]}{(1-x)^4} == \frac{e^x[(2-x)(1-x) + 3(3-x)]}{(1-x)^4} =$$

$$= \frac{e^x[2 - 2x - x + x^2 + 9 - 3x]}{(1-x)^4} = \frac{e^x(x^2 - 6x + 11)}{(1-x)^4}$$

The second derivative, at the point x where the first derivative is equal to 0, it is:

$$f''(3) = \frac{e^3(3^2 - 6\cdot 3 + 11)}{(1-3)^4} = \frac{e^3(9 - 18 + 11)}{(-2)^4} = \frac{2e^3}{16} = \frac{e^3}{8} > 0$$

$x = 3$ is a point of minimum.
The value of the function at the minimum point is:

$$f(3) = \frac{e^3}{(1-3)^2} = \frac{e^3}{4}$$

11.2.23.5. Inflection Points

We verify whether the function has inflection points, checking whether the second derivative is equal to 0.

$$f''(x) = \frac{e^x\left(x^2 - 6x + 11\right)}{(1-x)^4} = 0$$

The only solution is that for which the numerator is equal to 0:

$$x^2 - 6x + 11 = 0$$

The roots of second degree equation are:

$$x = \frac{-b \pm \sqrt{b^2 - 4ac}}{2a} = \frac{6 \pm \sqrt{36-44}}{2} = \frac{6 \pm \sqrt{-8}}{2}$$

The square root of a negative number does not admit real solutions, so there are no inflection points.

The graph of the function is shown in Figure 11.23.

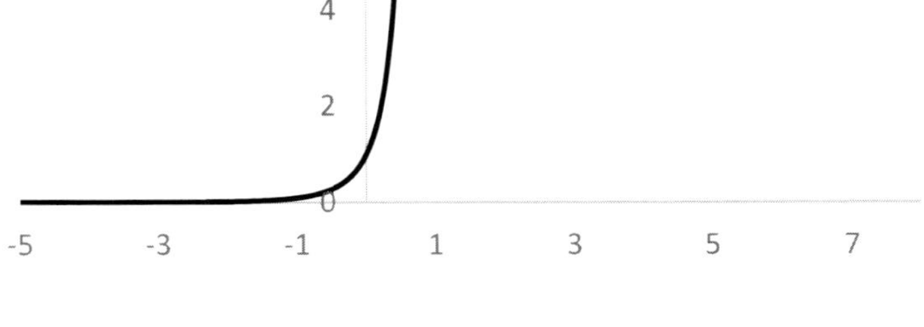

Figure 11.23. Graph of the function $f(x) = \dfrac{e^x}{(1-x)^2}$.

11.2.24. Problem 11.24

Study the function:

$$f(x) = \frac{e^{-x}}{(x-2)^2}$$

11.2.24.1. Domain of the Function

The domain of the function is given by the intersection of the function domains in the numerator and the denominator, in this case ***R***, minus the values in which the denominator is equal to 0.

The domain of the function is therefore determined by the values for which the denominator is not equal to 0:

$$x - 2 \neq 0 \quad \Rightarrow \quad x \neq 2$$

The function is therefore defined by:

$$f : \boldsymbol{R} - \{2\} \to \boldsymbol{R}$$

11.2.24.2. Intersection with the Axes

We verify whether the function intersects the x, y axes:

$$x = 0 \quad \to \quad f(0) = ?$$
$$y = f(x) = 0 \quad \to \quad x = ?$$

For $x = 0$:

$$y = f(0) = \frac{e^0}{(0-2)^2} = \frac{1}{4}$$

The function intersects the y-axis at the point $y = \frac{1}{4}$.

For $y = 0$:

$$y = f(x) = \frac{e^{-x}}{(x-2)^2} = 0$$

The equation does not admit a solution, so the function does not intersect the x-axis.

11.2.24.3. Asymptotes

There is a vertical asymptote if:

$$\lim_{x \to x_0 +} \frac{e^{-x}}{(x-2)^2} = \pm\infty$$

$$\lim_{x \to x_0 -} \frac{e^{-x}}{(x-2)^2} = \pm\infty$$

Given the function definition domain, we can check the limits for $x_0 = 2$:

$$\lim_{x \to 2-} \frac{e^{-x}}{(x-2)^2} = \frac{e^{-2}}{0^+} = +\infty$$

$$\lim_{x \to 2+} \frac{e^{-x}}{(x-2)^2} = \frac{e^{-2}}{0^+} = +\infty$$

There is a vertical asymptote at $x = 2$ with two infinite branches to $+\infty$.
There is a horizontal asymptote if:

$$\lim_{x \to -\infty} \frac{e^{-x}}{(x-2)^2} = l_1 \ , \ l_1 \in \boldsymbol{R}$$

$$\lim_{x \to +\infty} \frac{e^{-x}}{(x-2)^2} = l_2 \ , \ l_2 \in \boldsymbol{R}$$

We have:

$$\lim_{x \to +\infty} \frac{e^{-x}}{(x-2)^2} = \frac{e^{-\infty}}{(+\infty - 2)^2} = \frac{0}{+\infty} = 0$$

For $x \to +\infty$ there is a horizontal asymptote at $y = 0$.
For $x \to -\infty$ we have:

$$\lim_{x\to-\infty}\frac{e^{-x}}{(x-2)^2}=\frac{e^{+\infty}}{(-\infty-2)^2}=\frac{+\infty}{+\infty}$$

It is an indeterminate form, to solve it we can apply De L'Hôpital's rule:

$$\lim_{x\to-\infty}\frac{e^{-x}}{(x-2)^2}=\lim_{x\to-\infty}\frac{\frac{d}{dx}\left(e^{-x}\right)}{\frac{d}{dx}(x-2)^2}=\lim_{x\to-\infty}\frac{-e^{-x}}{2(x-2)}=\frac{-\infty}{-\infty}$$

By continuing to apply De L'Hôpital's rule we have:

$$\lim_{x\to-\infty}\frac{-e^{-x}}{2(x-2)}=\lim_{x\to-\infty}\frac{\frac{d}{dx}\left(-e^{-x}\right)}{\frac{d}{dx}[2(x-2)]}=\lim_{x\to-\infty}\frac{e^{-x}}{2}=+\infty$$

There is no horizontal asymptote for $x\to-\infty$, so it is necessary to check whether there is an oblique asymptote:

$$\lim_{x\to-\infty}\frac{f(x)}{x}=\lim_{x\to-\infty}\frac{\frac{e^{-x}}{(x-2)^2}}{x}=\lim_{x\to-\infty}\frac{e^{-x}}{x(x-2)^2}=m,\ m\in \boldsymbol{R}$$

We have:

$$\lim_{x\to-\infty}\frac{e^{-x}}{x(x-2)^2}=\frac{e^{+\infty}}{-\infty\cdot(-\infty-2)^2}=\frac{+\infty}{(-\infty)(+\infty)}$$

It is an indeterminate form. We can solve the indeterminate form and calculate the limit by applying De L'Hôpital's rule.

$$\lim_{x\to-\infty}\frac{e^{-x}}{x(x-2)^2}=\lim_{x\to-\infty}\frac{\frac{d}{dx}\left(e^{-x}\right)}{\frac{d}{dx}\left[x(x-2)^2\right]}=$$

$$=\lim_{x\to-\infty}\frac{-e^{-x}}{(x-2)^2+2x(x-2)}=\frac{-\infty}{+\infty+\infty+\infty}$$

It is still an indeterminate form, we can continue to apply De L'Hôpital's rule:

$$\lim_{x\to-\infty}\frac{-e^{-x}}{(x-2)^2+2x(x-2)}=\lim_{x\to-\infty}\frac{\frac{d}{dx}\left(-e^{-x}\right)}{\frac{d}{dx}\left[(x-2)^2+2x(x-2)\right]}=$$

$$=\lim_{x\to-\infty}\frac{e^{-x}}{2(x-2)+2(x-2)+2x}=\lim_{x\to-\infty}\frac{e^{-x}}{2x-4+2x-4+2x}=$$

$$=\lim_{x\to-\infty}\frac{e^{-x}}{6x-8}=\frac{+\infty}{-\infty}$$

By continuing to apply De L'Hôpital's rule we have:

$$\lim_{x\to-\infty}\frac{e^{-x}}{6x-8}=\lim_{x\to-\infty}\frac{\frac{d}{dx}\left(e^{-x}\right)}{\frac{d}{dx}(6x-8)}=\lim_{x\to-\infty}\frac{-e^{-x}}{6}=-\infty$$

There is no oblique asymptote for $x\to-\infty$.

11.2.24.4. Maximum and/or Minimum Point

We verify if the function has maximum or minimum points, checking if the first derivative is equal to 0 at some point and if for these points the second derivative is > 0 or < 0.

$$f'(x_i)=0$$

$$f''(x_i)>0 \quad or \quad f''(x_i)<0$$

First derivative:

$$f'(x)=\frac{d}{dx}\left(\frac{e^{-x}}{(x-2)^2}\right)=\frac{-e^{-x}\cdot(x-2)^2-e^{-x}\cdot 2(x-2)}{(x-2)^4}=$$

$$=\frac{(x-2)\cdot\left[-e^{-x}(x-2)-2e^{-x}\right]}{(x-2)^4}=\frac{-e^{-x}(x-2+2)}{(x-2)^3}=\frac{-xe^{-x}}{(x-2)^3}$$

$$f'(x) = \frac{-xe^{-x}}{(x-2)^3} = 0$$

The only solution is that for which the numerator is equal to 0. The first derivative is equal to 0 at the point $x = 0$.

Second derivative:

$$f''(x) = \frac{d}{dx}\left(\frac{-xe^{-x}}{(x-2)^3}\right) = \frac{(-e^{-x} + xe^{-x})(x-2)^3 + xe^{-x} \cdot 3(x-2)^2}{(x-2)^6} =$$

$$= \frac{e^{-x}(x-2)^2[(-1+x)(x-2)+3x]}{(x-2)^6} = \frac{e^{-x}[-x+2+x^2-2x+3x]}{(x-2)^4} =$$

$$= \frac{e^{-x}(x^2+2)}{(x-2)^4}$$

The second derivative, at the point x where the first derivative is equal to 0, it is:

$$f''(0) = \frac{e^0(0^2+2)}{(0-2)^4} = \frac{2e^0}{(-2)^4} = \frac{2}{16} = \frac{1}{8} > 0$$

$x = 0$ is a point of minimum.

The value of the function at the minimum point is:

$$f(0) = \frac{e^0}{(0-2)^2} = \frac{1}{4}$$

11.2.24.5. Inflection Points

We verify whether the function has inflection points, checking whether the second derivative is equal to 0.

$$f''(x) = \frac{e^{-x}(x^2+2)}{(x-2)^4} = 0$$

The only solution is that for which the numerator is equal to 0:

$$x^2 + 2 = 0$$

The roots of second degree equation are:

$$x = \pm\sqrt{-2}$$

The square root of a negative number does not admit real solutions, so there are no inflection points.

The graph of the function is shown in Figure 11.24.

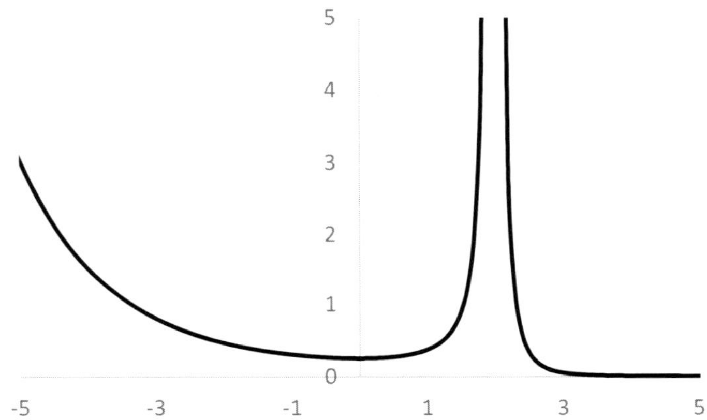

Figure 11.24. Graph of the function $f(x) = \dfrac{e^{-x}}{(x-2)^2}$.

11.1.25. Problem 11.25

Study the function:

$$f(x) = \frac{-e^{-x}}{(1+x)^2}$$

11.2.25.1. Domain of the Function

The domain of the function is given by the intersection of the function domains in the numerator and the denominator, in this case ***R***, minus the values in which the denominator is equal to 0.

The domain of the function is therefore determined by the values for which the denominator is not equal to 0:

$$1+x \neq 0 \quad \Rightarrow \quad x \neq -1$$

The function is therefore defined by:

$$f : \boldsymbol{R} - \{-1\} \rightarrow \boldsymbol{R}$$

11.2.25.2. Intersection with the Axes

We verify whether the function intersects the *x*, *y* axes:

$$x = 0 \quad \rightarrow \quad f(0) = ?$$

$$y = f(x) = 0 \quad \rightarrow \quad x = ?$$

For $x = 0$:

$$y = f(0) = \frac{-e^0}{(1+0)^2} = -1$$

The function intersects the *y*-axis at the point $y = -1$.

For $y = 0$:

$$y = f(x) = \frac{-e^{-x}}{(1+x)^2} = 0$$

The equation does not admit a solution, so the function does not intersect the *x*-axis.

11.2.25.3. Asymptotes

There is a vertical asymptote if:

$$\lim_{x \to x_0+} \frac{-e^{-x}}{(1+x)^2} = \pm\infty$$

$$\lim_{x \to x_0-} \frac{-e^{-x}}{(1+x)^2} = \pm\infty$$

Given the domain of the function, let's check the limits for $x_0 = -1$:

$$\lim_{x\to -1-} \frac{-e^{-x}}{(1+x)^2} = \frac{-e^1}{0^+} = -\infty$$

$$\lim_{x\to -1+} \frac{-e^{-x}}{(1+x)^2} = \frac{-e^1}{0^+} = -\infty$$

Where 0^+ is determined by the square of $(1 + x)$.

There is a vertical asymptote at $x = -1$ with two infinite branches to $-\infty$.

There is a horizontal asymptote if:

$$\lim_{x\to -\infty} \frac{-e^{-x}}{(1+x)^2} = l_1 \ , \ l_1 \in \boldsymbol{R}$$

$$\lim_{x\to +\infty} \frac{-e^{-x}}{(1+x)^2} = l_2 \ , \ l_2 \in \boldsymbol{R}$$

We have:

$$\lim_{x\to +\infty} \frac{-e^{-x}}{(1+x)^2} = \frac{-e^{-\infty}}{(1+\infty)^2} = \frac{0}{+\infty} = 0$$

For $x \to +\infty$ there is a horizontal asymptote at $y = 0$.

For $x \to -\infty$ we have:

$$\lim_{x\to -\infty} \frac{-e^{-x}}{(1+x)^2} = \frac{-e^{+\infty}}{(1-\infty)^2} = \frac{-\infty}{+\infty}$$

It is an indeterminate form that can be solved and calculated by applying De L'Hôpital's rule:

$$\lim_{x\to -\infty} \frac{-e^{-x}}{(1+x)^2} = \lim_{x\to -\infty} \frac{\frac{d}{dx}\left(-e^{-x}\right)}{\frac{d}{dx}(1+x)^2} = \lim_{x\to -\infty} \frac{e^{-x}}{2(1+x)} = \frac{+\infty}{-\infty}$$

By continuing to apply De L'Hôpital's rule we have:

$$\lim_{x\to-\infty}\frac{e^{-x}}{2(1+x)}=\lim_{x\to-\infty}\frac{\frac{d}{dx}\left(e^{-x}\right)}{\frac{d}{dx}[2(1+x)]}=\lim_{x\to-\infty}\frac{-e^{-x}}{2}=-\infty$$

There is no horizontal asymptote for $x\longrightarrow-\infty$, so it is necessary to check whether there is an oblique asymptote:

$$\lim_{x\to-\infty}\frac{f(x)}{x}=\lim_{x\to-\infty}\frac{\frac{-e^{-x}}{(1+x)^2}}{x}=\lim_{x\to-\infty}\frac{-e^{-x}}{x(1+x)^2}=m,\ m\in\boldsymbol{R}$$

We have:

$$\lim_{x\to-\infty}\frac{-e^{-x}}{x(1+x)^2}=\frac{-e^{+\infty}}{-\infty\cdot(1-\infty)^2}=\frac{-\infty}{(-\infty)(+\infty)}$$

It is an indeterminate form. We can solve the indeterminate form and calculate the limit by applying De L'Hôpital's rule:

$$\lim_{x\to-\infty}\frac{-e^{-x}}{x(1+x)^2}=\lim_{x\to-\infty}\frac{\frac{d}{dx}\left(-e^{-x}\right)}{\frac{d}{dx}\left[x(1+x)^2\right]}=\lim_{x\to-\infty}\frac{e^{-x}}{(1+x)^2+2x(1+x)}=$$

$$=\lim_{x\to-\infty}\frac{e^{-x}}{1+2x+x^2+2x+2x^2}=\lim_{x\to-\infty}\frac{e^{-x}}{3x^2+4x+1}=\frac{+\infty}{+\infty-\infty}$$

It is still an indeterminate form, we can continue to apply De L'Hôpital's rule:

$$\lim_{x\to-\infty}\frac{e^{-x}}{3x^2+4x+1}=\lim_{x\to-\infty}\frac{\frac{d}{dx}\left(e^{-x}\right)}{\frac{d}{dx}\left(3x^2+4x+1\right)}=\lim_{x\to-\infty}\frac{-e^{-x}}{6x+4}=\frac{-\infty}{-\infty}$$

$$\lim_{x\to-\infty}\frac{-e^{-x}}{6x+4}=\lim_{x\to-\infty}\frac{\frac{d}{dx}\left(-e^{-x}\right)}{\frac{d}{dx}(6x-8)}=\lim_{x\to-\infty}\frac{e^{-x}}{6}=+\infty$$

There is no oblique asymptote for $x \to -\infty$.

11.2.25.4. Maximum and/or Minimum Points

We verify if the function has maximum or minimum points, checking if the first derivative is equal to 0 at some point and if for these points the second derivative is > 0 or < 0.

$$f'(x_i)=0$$
$$f''(x_i)>0 \quad or \quad f''(x_i)<0$$

First derivative:

$$f'(x)=\frac{d}{dx}\left(\frac{-e^{-x}}{(1+x)^2}\right)=\frac{e^{-x}\cdot(1+x)^2-\left(-e^{-x}\right)\cdot 2(1+x)}{(1+x)^4}=$$
$$=\frac{(1+x)\cdot\left[e^{-x}(1+x)+2e^{-x}\right]}{(1+x)^4}=\frac{e^{-x}(x+3)}{(1+x)^3}$$

$$f'(x)=\frac{e^{-x}(x+3)}{(1+x)^3}=0$$

The only solution is that for which the numerator is equal to 0:

$$x+3=0 \quad \Rightarrow \quad x=-3$$

The first derivative is equal to 0 at the point $x = -3$.

Second derivative:

$$f''(x) = \frac{d}{dx}\left(\frac{e^{-x}(x+3)}{(1+x)^3}\right) =$$

$$= \frac{\left[-e^{-x}(x+3)+e^{-x}\right](1+x)^3 - e^{-x}(x+3)\cdot 3(1+x)^2}{(1+x)^6} =$$

$$= \frac{-e^{-x}(1+x)^2\left[(x+3-1)(1+x)+3(x+3)\right]}{(1+x)^6} =$$

$$= \frac{-e^{-x}\left[(x+2)(1+x)+3(x+3)\right]}{(1+x)^4} = \frac{-e^{-x}\left[x+x^2+2+2x+3x+9\right]}{(1+x)^4} =$$

$$= \frac{-e^{-x}\left(x^2+6x+11\right)}{(1+x)^4}$$

The second derivative, at the point x where the first derivative is equal to 0, it is:

$$f''(-3) = \frac{-e^3\left[(-3)^2+6\cdot(-3)+11\right]}{(1-3)^4} = \frac{-e^3(9-18+11)}{(-2)^4} =$$

$$= \frac{-2e^3}{16} = -\frac{e^3}{8} < 0$$

$x = -3$ is a point of maximum.

The value of the function at the maximum point is:

$$f(-3) = \frac{-e^3}{(1-3)^2} = -\frac{e^3}{4}$$

11.2.25.5. Inflection Points

We verify whether the function has inflection points, checking whether the second derivative is equal to 0.

$$f''(x) = \frac{-e^{-x}\left(x^2+6x+11\right)}{(1+x)^4} = 0$$

The only solution is that for which the numerator is equal to 0:

$$x^2+6x+11=0$$

The roots of second degree equation are:

$$x = \frac{-b \pm \sqrt{b^2 - 4ac}}{2a} = \frac{-6 \pm \sqrt{36 - 44}}{2} = \frac{-6 \pm \sqrt{-8}}{2}$$

The square root of a negative number does not admit real solutions, so there are no inflection points.

The graph of the function is shown in Figure 11.25.

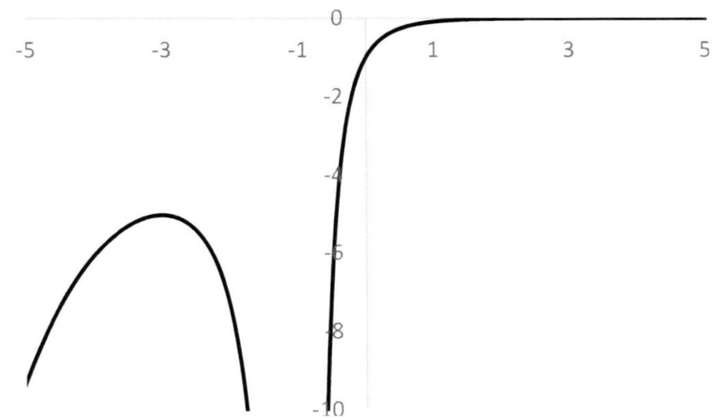

Figure 11.25. Graph of the function $f(x) = \frac{-e^{-x}}{(1+x)^2}$.

11.2.26. Problem 11.26

Study the function:

$$f(x) = \frac{-e^x}{(3-x)^2}$$

11.2.26.1. Domain of the Function

The domain of the function is given by the intersection of the function domains in the numerator and the denominator, in this case ***R***, minus the values in which the denominator is equal to 0.

The domain of the function is therefore determined by the values for which the denominator is not equal to 0:

$$3 - x \neq 0 \quad \Rightarrow \quad x \neq 3$$

The function is therefore defined by:

$$f : \boldsymbol{R} - \{3\} \to \boldsymbol{R}$$

11.2.26.2. Intersection with the Axes

We verify whether the function intersects the x, y axes:

$$x = 0 \quad \to \quad f(0) = ?$$

$$y = f(x) = 0 \quad \to \quad x = ?$$

For $x = 0$:

$$y = f(0) = \frac{-e^0}{(3-0)^2} = -\frac{1}{9}$$

The function intersects the y-axis at the point $y = -\frac{1}{9}$.

For $y = 0$:

$$y = f(x) = \frac{-e^x}{(3-x)^2} = 0$$

The equation does not admit a solution, so the function does not intersect the x-axis.

11.2.26.3. Asymptotes

There is a vertical asymptote if:

$$\lim_{x \to x_0 +} \frac{-e^x}{(3-x)^2} = \pm\infty$$

$$\lim_{x \to x_0 -} \frac{-e^x}{(3-x)^2} = \pm\infty$$

Given the domain of the function, let's check the limits for $x_0 = 3$:

$$\lim_{x\to 3-}\frac{-e^x}{(3-x)^2}=\frac{-e^3}{0^+}=-\infty$$

$$\lim_{x\to 3+}\frac{-e^x}{(3-x)^2}=\frac{-e^3}{0^+}=-\infty$$

Where 0^+ is determined by the square of $(3-x)$.

There is a vertical asymptote at $x_0 = 3$ with two infinite branches to $-\infty$.

There is a horizontal asymptote if:

$$\lim_{x\to -\infty}\frac{-e^x}{(3-x)^2}=l_1\ ,\ l_1\in \boldsymbol{R}$$

$$\lim_{x\to +\infty}\frac{-e^x}{(3-x)^2}=l_2\ ,\ l_2\in \boldsymbol{R}$$

We have:

$$\lim_{x\to -\infty}\frac{-e^x}{(3-x)^2}=\frac{-e^{-\infty}}{(3+\infty)^2}=\frac{0}{+\infty}=0$$

For $x \to -\infty$ there is a horizontal asymptote at $y = 0$.

For $x \to +\infty$ we have:

$$\lim_{x\to +\infty}\frac{-e^x}{(3-x)^2}=\frac{-e^{+\infty}}{(1-\infty)^2}=\frac{-\infty}{+\infty}$$

It is an indeterminate form that can be solved and calculated by applying De L'Hôpital's rule:

$$\lim_{x\to +\infty}\frac{-e^x}{(3-x)^2}=\lim_{x\to +\infty}\frac{\frac{d}{dx}\left(-e^x\right)}{\frac{d}{dx}(3-x)^2}=\lim_{x\to +\infty}\frac{-e^x}{-2(3-x)}=\frac{-\infty}{+\infty}$$

By continuing to apply De L'Hôpital's rule we have:

$$\lim_{x\to +\infty}\frac{-e^x}{-2(3-x)}=\lim_{x\to +\infty}\frac{\frac{d}{dx}\left(-e^x\right)}{\frac{d}{dx}\left[-2(3-x)\right]}=\lim_{x\to +\infty}\frac{-e^x}{2}=-\infty$$

There is no horizontal asymptote for $x \to +\infty$, so it is necessary to check whether there is an oblique asymptote:

$$\lim_{x\to+\infty} \frac{f(x)}{x} = \lim_{x\to+\infty} \frac{\frac{-e^x}{(3-x)^2}}{x} = \lim_{x\to+\infty} \frac{-e^x}{x(3-x)^2} = m\,,\ m \in \boldsymbol{R}$$

We have:

$$\lim_{x\to+\infty} \frac{-e^x}{x(3-x)^2} = \frac{-e^{+\infty}}{+\infty\cdot(1-\infty)^2} = \frac{-\infty}{+\infty}$$

We can solve the indeterminate form and calculate the limit by applying De L'Hôpital's rule:

$$\lim_{x\to+\infty} \frac{-e^x}{x(3-x)^2} = \lim_{x\to+\infty} \frac{\frac{d}{dx}\left(-e^x\right)}{\frac{d}{dx}\left[x(3-x)^2\right]} = \lim_{x\to+\infty} \frac{-e^x}{(3-x)^2-2x(3-x)} =$$

$$= \lim_{x\to+\infty} \frac{-e^x}{9+x^2-6x-6x+2x^2} = \lim_{x\to+\infty} \frac{-e^x}{3x^2-12x+9} = \frac{-\infty}{+\infty-\infty}$$

It is still an indeterminate form, we can continue to apply De L'Hôpital's rule:

$$\lim_{x\to+\infty} \frac{-e^x}{3x^2-12x+9} = \lim_{x\to+\infty} \frac{\frac{d}{dx}\left(-e^x\right)}{\frac{d}{dx}\left(3x^2-12x+9\right)} =$$

$$= \lim_{x\to+\infty} \frac{-e^x}{6x-12} = \frac{-\infty}{+\infty}$$

$$\lim_{x\to+\infty} \frac{-e^x}{6x-12} = \lim_{x\to+\infty} \frac{\frac{d}{dx}\left(-e^x\right)}{\frac{d}{dx}(6x-12)} = \lim_{x\to+\infty} \frac{-e^x}{6} = -\infty$$

There is no oblique asymptote for $x \to +\infty$.

11.2.26.4. Maximum and/or Minimum Points

We verify if the function has maximum or minimum points, checking if the first derivative is equal to 0 at some point and if for these points the second derivative is > 0 or < 0.

$$f'(x_i) = 0$$

$$f''(x_i) > 0 \qquad or \qquad f''(x_i) < 0$$

First derivative:

$$f'(x) = \frac{d}{dx}\left(\frac{-e^x}{(3-x)^2}\right) = \frac{-e^x \cdot (3-x)^2 - (-e^x)(-2)(3-x)}{(3-x)^4} =$$

$$= \frac{-e^x(3-x)\cdot(3-x+2)}{(3-x)^4} = \frac{-e^x(5-x)}{(3-x)^3}$$

$$f'(x) = \frac{-e^x(5-x)}{(3-x)^3} = 0$$

The only solution is that for which the numerator is equal to 0:

$$5 - x = 0 \quad \Rightarrow \quad x = 5$$

The first derivative is equal to 0 at the point $x = 5$.
Second derivative:

$$f''(x) = \frac{d}{dx}\left(\frac{-e^x(5-x)}{(3-x)^3}\right) =$$

$$= \frac{[-e^x(5-x) - e^x(-1)](3-x)^3 - [-e^x(5-x)](-3)(3-x)^2}{(3-x)^6} =$$

$$= \frac{(3-x)^2[(-5e^x + xe^x + e^x)(3-x) - 3e^x(5-x)]}{(3-x)^6} =$$

$$= \frac{[e^x(x-4)(3-x) - 3e^x(5-x)]}{(3-x)^4} = \frac{e^x[(x-4)(3-x) - 3(5-x)]}{(3-x)^4} =$$

$$= \frac{e^x[3x - x^2 - 12 + 4x - 15 + 3x]}{(3-x)^4} = \frac{e^x(-x^2 + 10x - 27)}{(3-x)^4}$$

The second derivative, at the point x where the first derivative is equal to 0, it is:

$$f''(5) = \frac{e^5(-5^2 + 10 \cdot 5 - 27)}{(3-5)^4} = \frac{e^5(-25 + 50 - 27)}{(-2)^4} =$$

$$= \frac{-2e^5}{16} = -\frac{e^5}{8} < 0$$

$x = 5$ is a point of maximum.
The value of the function at the maximum point is:

$$f(5) = \frac{-e^5}{(3-5)^2} = -\frac{e^5}{4}$$

11.2.26.5. Inflection Points

We verify whether the function has inflection points, checking whether the second derivative is equal to 0.

$$f''(x) = \frac{e^x(-x^2 + 10x - 27)}{(3-x)^4} = 0$$

The only solution is that for which the numerator is equal to:

$$-x^2 + 10x - 27 = 0$$

The roots of second degree equation are:

$$x = \frac{-b \pm \sqrt{b^2 - 4ac}}{2a} = \frac{-10 \pm \sqrt{100 - 108}}{-2} = \frac{-10 \pm \sqrt{-8}}{-2}$$

The square root of a negative number does not admit real solutions, so there are no inflection points

The graph of the function is shown in Figure 11.26.

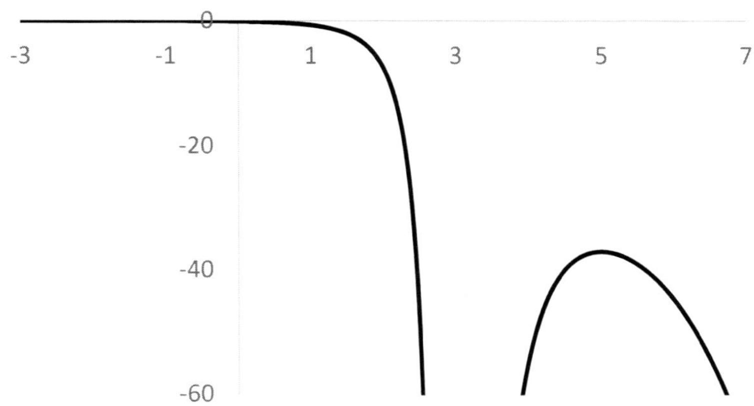

Figure 11.26. Graph of the function $f(x)=\dfrac{-e^{x}}{(3-x)^{2}}$.

11.1.27. Problem 11.27

Study the function:

$$f(x)=\frac{2\sqrt{x}-1}{x}$$

11.2.27.1. Domain of the Function

The domain of the function is given by the intersection of the function domains in the numerator and the denominator, minus the values in which the denominator is equal to 0.

The denominator domain is ***R***, so the function definition domain is the numerator domain, in this case $[0,+\infty[$, minus the values where the denominator equals 0:

$$x=0$$

The function is therefore defined by:

$$f:\boldsymbol{R}^{+}\rightarrow\boldsymbol{R}$$

11.2.27.2. Intersection with the Axes

We verify whether the function intersects the x, y axes:

$$x=0 \quad \rightarrow \quad f(0)=?$$
$$y=f(x)=0 \quad \rightarrow \quad x=?$$

The value $x = 0$ does not belong to the domain of the function, so the function does not intersect the y-axis.

For $y = 0$:

$$y = f(x) = \frac{2\sqrt{x} - 1}{x} = 0$$

The only solution is that for which the numerator is equal to 0:

$$2\sqrt{x} - 1 = 0 \quad \Rightarrow \quad 2\sqrt{x} = 1 \quad \Rightarrow \quad \sqrt{x} = \frac{1}{2} \quad \Rightarrow \quad x = \frac{1}{4}$$

The function intersects the x-axis at the point $\left(\frac{1}{4}, 0\right)$.

11.2.27.3. Asymptotes

There is a vertical asymptote if:

$$\lim_{x \to x_0+} \frac{2\sqrt{x} - 1}{x} = \pm\infty$$

$$\lim_{x \to x_0-} \frac{2\sqrt{x} - 1}{x} = \pm\infty$$

Given the function definition domain, we can check the limit for $x_0^+ = 0^+$:

$$\lim_{x \to 0+} \frac{2\sqrt{x} - 1}{x} = \frac{-1}{0} = -\infty$$

There is a vertical asymptote at $x_0 = 0$ with one infinite branch to $-\infty$.

There is a horizontal asymptote if:

$$\lim_{x \to -\infty} \frac{2\sqrt{x} - 1}{x} = l_1 \quad l_1 \in \boldsymbol{R}$$

$$\lim_{x \to +\infty} \frac{2\sqrt{x} - 1}{x} = l_2 \quad l_2 \in \boldsymbol{R}$$

Given the function definition domain, for horizontal asymptotes we check the limit only for $x \to +\infty$.

For the function it holds:

$$\lim_{x \to +\infty} \frac{2\sqrt{x} - 1}{x} = \frac{+\infty}{+\infty}$$

It is an indeterminate form, to solve it we can apply De L'Hôpital's rule:

$$\lim_{x \to +\infty} \frac{2\sqrt{x} - 1}{x} = \lim_{x \to +\infty} \frac{\frac{d}{dx}\left(2\sqrt{x} - 1\right)}{\frac{d}{dx}x} = \lim_{x \to +\infty} \frac{2 \cdot \frac{1}{2} x^{-\frac{1}{2}}}{1} =$$

$$= \lim_{x \to +\infty} \frac{1}{x^{\frac{1}{2}}} = \frac{1}{+\infty} = 0$$

For $x \to +\infty$ there is a horizontal asymptote at $y = 0$.

Since the function is continuous on the definition domain ($\boldsymbol{R}^+$) and given the existence of a vertical and a horizontal asymptote, there can be no oblique asymptotes.

11.2.27.4. Maximum and/or Minimum Points

We verify if the function has maximum or minimum points, checking if the first derivative is equal to 0 at some point and if for these points the second derivative is > 0 or < 0.

$$f'(x_i) = 0$$

$$f''(x_i) > 0 \quad or \quad f''(x_i) < 0$$

First derivative:

$$f'(x)=\frac{d}{dx}\left(\frac{2\sqrt{x}-1}{x}\right)=\frac{2\cdot\frac{1}{2}x^{-\frac{1}{2}}\cdot x-\left(2\sqrt{x}-1\right)\cdot 1}{x^2}=$$

$$=\frac{x^{\frac{1}{2}}-2x^{\frac{1}{2}}+1}{x^2}=\frac{-x^{\frac{1}{2}}+1}{x^2}$$

$$f'(x)=\frac{-x^{\frac{1}{2}}+1}{x^2}=0$$

The only solution is that for which the numerator is equal to 0:

$$-x^{\frac{1}{2}}+1=0 \quad\Rightarrow\quad x^{\frac{1}{2}}=1 \quad\Rightarrow\quad x=1$$

The first derivative is equal to 0 at the point $x = 1$.
Second derivative:

$$f''(x)=\frac{d}{dx}\left(\frac{-x^{\frac{1}{2}}+1}{x^2}\right)=\frac{-\frac{1}{2}x^{-\frac{1}{2}}\cdot x^2-\left(-x^{\frac{1}{2}}+1\right)\cdot 2x}{x^4}=$$

$$=\frac{-\frac{1}{2}x^{\frac{3}{2}}+2x^{\frac{3}{2}}-2x}{x^4}=\frac{\frac{3}{2}x^{\frac{3}{2}}-2x}{x^4}=\frac{\frac{3}{2}x^{\frac{1}{2}}-2}{x^3}$$

The second derivative, at the point x where the first derivative is equal to 0, it is:

$$f''(1)=\frac{\frac{3}{2}-2}{1}=\frac{3-4}{2}=-\frac{1}{2}<0$$

$x = 1$ is a point of maximum.
The value of the function at the maximum point is:

$$f(1)=\frac{2-1}{1}=1$$

11.2.27.5. Inflection Points

We verify whether the function has inflection points, checking whether the second derivative is equal to 0.

$$f''(x) = \frac{\frac{3}{2}x^{\frac{1}{2}} - 2}{x^3} = 0$$

The only solution is that for which the numerator is equal to 0:

$$\frac{3}{2}x^{\frac{1}{2}} - 2 = 0 \Rightarrow \frac{3}{2}x^{\frac{1}{2}} = 2 \Rightarrow x^{\frac{1}{2}} = 2\cdot\frac{2}{3} = \frac{4}{3} \Rightarrow x = \frac{16}{9}$$

The second derivative is equal to 0 at the point $x = \frac{16}{9}$.

The value of the function at the inflection point is:

$$f(\frac{16}{9}) = \frac{2\sqrt{\frac{16}{9}} - 1}{\frac{16}{9}} = \frac{2\cdot\frac{4}{3} - 1}{\frac{16}{9}} = \frac{\frac{8}{3} - 1}{\frac{16}{9}} = \frac{\frac{8-3}{3}}{\frac{16}{9}} = \frac{\frac{5}{3}}{\frac{16}{9}} = \frac{5}{3}\cdot\frac{9}{16} = \frac{15}{16}$$

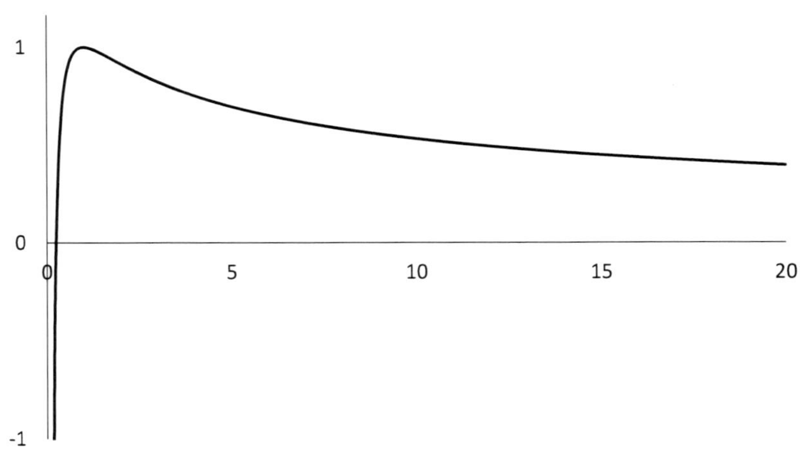

Figure 11.27. Graph of the function $f(x) = \frac{2\sqrt{x} - 1}{x}$.

The graph of the function is shown in Figure 11.27.

11.2.28. Problem 11.28

Study the function:

$$f(x) = \left(e^{x} + e^{-x}\right)^{3}$$

11.2.28.1. Domain of the Function

The function shows the sum of two exponentials, so the domain is ***R***:
The function is therefore defined by:

$$f : \mathbf{R} \rightarrow \mathbf{R}^{+}$$

11.2.28.2. Intersection with the Axes

We verify whether the function intersects the x, y axes:

$$x = 0 \quad \rightarrow \quad f(0) = ?$$
$$y = f(x) = 0 \quad \rightarrow \quad x = ?$$

For $x = 0$:

$$y = f(0) = \left(e^{0} + e^{-0}\right)^{3} = (1+1)^{3} = 2^{3} = 8$$

The function intersects the y-axis at the point $y = 8$.
For $y = 0$:

$$y = f(x) = \left(e^{x} + e^{-x}\right)^{3} = 0$$

It should be:

$$e^{x} + e^{-x} = 0$$

An exponential function is always > 0, so the equation has no solution and the function does not intersect the x-axis.

11.2.28.3. Asymptotes

There is a vertical asymptote if:

$$\lim_{x \to x_0 +} \left(e^x + e^{-x}\right)^3 = \pm\infty$$

$$\lim_{x \to x_0 -} \left(e^x + e^{-x}\right)^3 = \pm\infty$$

The function is continuous on $\boldsymbol{R}$, so there are no discontinuity points x_0 and the function has no vertical asymptotes.

There is a horizontal asymptote if:

$$\lim_{x \to +\infty} \left(e^x + e^{-x}\right)^3 = l_1 \ , \ l_1 \in \boldsymbol{R}$$

$$\lim_{x \to -\infty} \left(e^x + e^{-x}\right)^3 = l_2 \ , \ l_2 \in \boldsymbol{R}$$

We have:

$$\lim_{x \to +\infty} \left(e^x + e^{-x}\right)^3 = \left(e^{+\infty} + e^{-\infty}\right)^3 = (+\infty + 0)^3 = +\infty$$

$$\lim_{x \to -\infty} \left(e^x + e^{-x}\right)^3 = \left(e^{-\infty} + e^{+\infty}\right)^3 = (0 + \infty)^3 = +\infty$$

The function has no horizontal asymptotes for $x \to +\infty$ and $x \to -\infty$.

Since there are no horizontal asymptotes, we have to verify if there are oblique asymptotes.

$$\lim_{x \to +\infty} \frac{f(x)}{x} = m_1 \ ; \ \lim_{x \to +\infty} [f(x) - m_1 x] = q_1 \ ; \ m_1, q_1 \in \boldsymbol{R}$$

$$\lim_{x \to -\infty} \frac{f(x)}{x} = m_2 \ ; \ \lim_{x \to -\infty} [f(x) - m_2 x] = q_2 \ ; \ m_2, q_2 \in \boldsymbol{R}$$

In the present case we have:

$$\lim_{x \to +\infty} \frac{f(x)}{x} = \lim_{x \to +\infty} \frac{\left(e^x + e^{-x}\right)^3}{x} = \frac{+\infty}{+\infty}$$

It is an indeterminate form. To solve the indeterminate form we can apply De L'Hôpital's rule:

$$\lim_{x\to+\infty}\frac{\left(e^x+e^{-x}\right)^3}{x}=\lim_{x\to+\infty}\frac{\frac{d}{dx}\left(e^x+e^{-x}\right)^3}{\frac{d}{dx}(x)}=$$

$$=\lim_{x\to+\infty}\frac{3\left(e^x+e^{-x}\right)^2\cdot\left(e^x-e^{-x}\right)}{1}=3(+\infty)(+\infty)=+\infty$$

There is no oblique asymptote for $x \to +\infty$.

For $x \to -\infty$ it holds:

$$\lim_{x\to-\infty}\frac{f(x)}{x}=\lim_{x\to-\infty}\frac{\left(e^x+e^{-x}\right)^3}{x}=\frac{+\infty}{-\infty}$$

Also in this case, to resolve the indeterminate form we proceed by applying De L'Hôpital's rule:

$$\lim_{x\to+\infty}\frac{\left(e^x+e^{-x}\right)^3}{x}=\lim_{x\to-\infty}\frac{\frac{d}{dx}\left(e^x+e^{-x}\right)^3}{\frac{d}{dx}(x)}=$$

$$=\lim_{x\to-\infty}\frac{3\left(e^x+e^{-x}\right)^2\cdot\left(e^x-e^{-x}\right)}{1}=3(+\infty)(-\infty)=-\infty$$

There is no oblique asymptote for $x \to -\infty$.

11.2.28.4. Maximum and/or Minimum Points

We verify if the function has maximum or minimum points, checking if the first derivative is equal to 0 at some point and if for these points the second derivative is > 0 or < 0.

$$f'(x_i)=0$$

$$f''(x_i)>0 \quad or \quad f''(x_i)<0$$

First derivative:

$$f'(x)=\frac{d}{dx}\left(e^{x}+e^{-x}\right)^{3}=3\left(e^{x}+e^{-x}\right)^{2}\cdot\left(e^{x}-e^{-x}\right)$$

$$f'(x)=3\left(e^{x}+e^{-x}\right)^{2}\cdot\left(e^{x}-e^{-x}\right)=0$$

The only solution is:

$$e^{x}-e^{-x}=0\Rightarrow e^{x}=e^{-x}\Rightarrow \log e^{x}=\log e^{-x}\Rightarrow$$

$$x=-x\Leftrightarrow x=0$$

The first derivative is equal to 0 at the point $x = 0$.
Second derivative:

$$f''(x)=\frac{d}{dx}\left(3\left(e^{x}+e^{-x}\right)^{2}\cdot\left(e^{x}-e^{-x}\right)\right)=$$

$$=3\cdot 2\left(e^{x}+e^{-x}\right)\cdot\left(e^{x}-e^{-x}\right)\cdot\left(e^{x}-e^{-x}\right)+3\left(e^{x}+e^{-x}\right)^{2}\cdot\left(e^{x}+e^{-x}\right)=$$

$$=6\left(e^{x}+e^{-x}\right)\cdot\left(e^{x}-e^{-x}\right)^{2}+3\left(e^{x}+e^{-x}\right)\cdot\left(e^{x}+e^{-x}\right)^{2}=$$

$$=3\left(e^{x}+e^{-x}\right)\cdot\left[2\left(e^{x}-e^{-x}\right)^{2}+\left(e^{x}+e^{-x}\right)^{2}\right]$$

The second derivative, at the point x where the first derivative is equal to 0, it is:

$$f''(0)=3\left(e^{0}+e^{-0}\right)\cdot\left[2\left(e^{0}-e^{-0}\right)^{2}+\left(e^{0}+e^{-0}\right)^{2}\right]=$$

$$=3(1+1)\cdot\left[2(1-1)^{2}+(1+1)^{2}\right]=3\cdot 2\cdot 4=24>0$$

$x = 0$ is a point of minimum.
The value of the function at the minimum point is:

$$f(0)=\left(e^{0}+e^{-0}\right)^{3}=2^{3}=8$$

11.2.28.5. Inflection Points

We verify whether the function has inflection points, checking whether the second derivative is equal to 0:

$$f''(x)=3\left(e^{x}+e^{-x}\right)\left[2\left(e^{x}-e^{-x}\right)^{2}+\left(e^{x}+e^{-x}\right)^{2}\right]=0$$

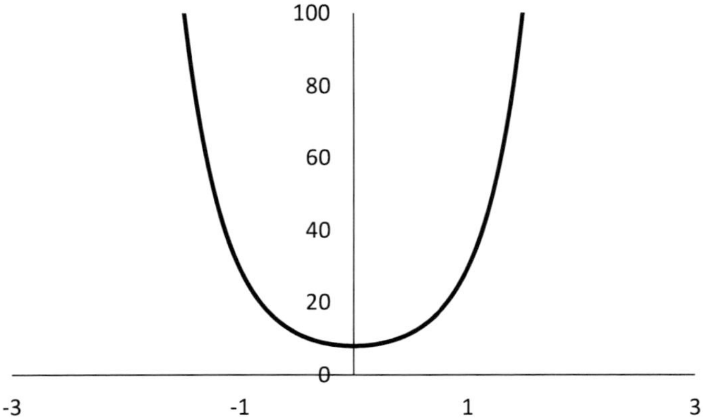

Figure 11.28. Graph of the function $f(x) = \left(e^{x} + e^{-x}\right)^{3}$.

Given the nature of the exponential function and given that only sums of positive values appear in the second derivative, the equation has no solution, so the function has no inflection points.

The graph of the function is shown in Figure 11.28.

11.2.29. Problem 11.29

Study the function:

$$f(x) = \frac{x}{1 - \log x}$$

11.2.29.1. Domain of the Function

The function in the numerator is defined on R, so the function definition domain depends on the denominator domain minus the values in which the denominator is equal to 0.

The logarithm function is defined by:

$$\log : \boldsymbol{R}^{+} \rightarrow \boldsymbol{R}$$

In this case we have the sum of a constant 1 and of the logarithm function, so the domain of the denominator is:

$$D = \left]0, +\infty\right[$$

The denominator equals 0 for:

$$1-\log x=0\Rightarrow \log x=1\Rightarrow e^{\log x}=e^{1}\Rightarrow x=e$$

value that must be excluded from the domain:

$$D=\left]0,e\right[\cup\left]e,+\infty\right[$$

The function is therefore defined by:

$$f:\left]0,e\right[\cup\left]e,+\infty\right[\rightarrow \boldsymbol{R}$$

11.2.29.2. Intersection with the Axes

We verify whether the function intersects the x, y axes:

$$x=0 \quad \rightarrow \quad f(0)=?$$

$$y=f(x)=0 \quad \rightarrow \quad x=?$$

The point $x = 0$ does not belong to the domain of the function, so the function does not intersect the y-axis.

For $y = 0$:

$$y=f(x)=\frac{x}{1-\log x}=0$$

The only solution would be that for which the numerator is equal to 0, but $x = 0$ is a value already excluded from the domain, so the function does not intersect the x-axis.

11.2.29.3. Asymptotes

There is a vertical asymptote if:

$$\lim_{x\rightarrow x_0+}\frac{x}{1-\log x}=\pm\infty$$

$$\lim_{x\rightarrow x_0-}\frac{x}{1-\log x}=\pm\infty$$

Given the domain of the function, let's check the limits for $x_0 = 0^+$, e^- and e^+:

$$\lim_{x \to 0+} \frac{x}{1 - \log x} = \frac{0^+}{1 - \log(0^+)} = \frac{0^+}{+\infty} = 0$$

there is no vertical asymptote.

$$\lim_{x \to e-} \frac{x}{1 - \log x} = \frac{e}{1 - 1^-} = \frac{e}{0^+} = +\infty$$

$$\lim_{x \to e+} \frac{x}{1 - \log x} = \frac{e}{1 - 1^+} = \frac{e}{0^-} = -\infty$$

There is a vertical asymptote at the point $x_0 = e$ with two infinite branches, $-\infty$ and $+\infty$.

Given the domain of the function, there is a horizontal asymptote if:

$$\lim_{x \to +\infty} \frac{x}{1 - \log x} = l_1 \ , \ l_1 \in \boldsymbol{R}$$

We have:

$$\lim_{x \to +\infty} \frac{x}{1 - \log x} = \frac{+\infty}{-\infty}$$

It is an indeterminate form, to solve it we can apply De L'Hôpital's rule:

$$\lim_{x \to +\infty} \frac{x}{1 - \log x} = \lim_{x \to +\infty} \frac{\frac{d}{dx}(x)}{\frac{d}{dx}(1 - \log x)} = \lim_{x \to +\infty} \frac{1}{-\frac{1}{x}} = \lim_{x \to +\infty} (-x) = -\infty$$

There is no horizontal asymptote for $x \to +\infty$, so it is necessary to check whether there is an oblique asymptote at the straight line $y = mx + q$. This is true if:

$$\lim_{x \to +\infty} \frac{f(x)}{x} = m \ ; \ \lim_{x \to +\infty} [f(x) - mx] = q \ ; \ m, q \in \boldsymbol{R}$$

For the coefficient m we obtain:

$$\lim_{x\to+\infty}\frac{f(x)}{x}=\lim_{x\to+\infty}\frac{1}{x}\cdot\frac{x}{1-\log(x+1)}=\lim_{x\to+\infty}\frac{1}{1-\log x}=\frac{1}{-\infty}=0$$

Since $m = 0$, the function does not have an oblique asymptote.

In fact $m = 0$ corresponds to a horizontal straight line, that is a horizontal asymptote that, if there was, it should have been identified previously.

11.2.29.4. Maximum and/or Minimum Points

We verify if the function has maximum or minimum points, checking if the first derivative is equal to 0 at some point and if for these points the second derivative is > 0 or < 0.

$$f'(x_i)=0$$
$$f''(x_i)>0 \quad or \quad f''(x_i)<0$$

First derivative:

$$f'(x)=\frac{d}{dx}\left(\frac{x}{1-\log x}\right)=\frac{1\cdot(1-\log x)-x\cdot\left(-\frac{1}{x}\right)}{(1-\log x)^2}=$$
$$=\frac{1-\log x+1}{(1-\log x)^2}=\frac{2-\log x}{(1-\log x)^2}$$
$$f'(x)=\frac{2-\log x}{(1-\log x)^2}=0$$

The only solution is that for which the numerator is equal to 0:

$$2-\log x=0\Rightarrow \log x=2\Rightarrow e^{\log x}=e^2\Rightarrow x=e^2$$

The first derivative is equal to 0 at the point $x = e^2$.

Second derivative:

$$f''(x) = \frac{d}{dx}\left(\frac{2-\log x}{(1-\log x)^2}\right) =$$

$$= \frac{-\frac{1}{x}\cdot(1-\log x)^2 - (2-\log x)\cdot 2(1-\log x)\cdot\left(-\frac{1}{x}\right)}{(1-\log x)^4} =$$

$$= \frac{-\frac{1}{x}\cdot(1-\log x)\cdot[-2(2-\log x)]}{(1-\log x)^4} = -\frac{-4+\log x}{x(1-\log x)^3} = \frac{4-\log x}{x(1-\log x)^3}$$

The second derivative, at the point where the first derivative is equal to 0, it is:

$$f''(e^2) = \frac{4-\log e^2}{e^2(1-\log e^2)^3} = \frac{4-2\log e}{e^2(1-2\log e)^3} = \frac{4-2}{e^2(1-2)^3} = \frac{2}{-e^2} < 0$$

$x = e^2$ is a point of maximum.
The value of the function at the maximum point is:

$$f(e^2) = \frac{e^2}{1-\log e^2} = \frac{e^2}{1-2\log e} = \frac{e^2}{1-2} = -e^2$$

11.2.29.5. Inflection Points

We verify whether the function has inflection points, checking whether the second derivative is equal to 0.

$$f''(x) = \frac{4-\log x}{x(1-\log x)^3} = 0$$

The only solution is that for which the numerator is equal to 0:

$$4-\log x = 0 \Rightarrow \log x = 4 \Rightarrow e^{\log x} = e^4 \Rightarrow x = e^4$$

The second derivative is equal to 0 at the point $x = e^2 - 1$.
The value of the function at the inflection point is:

$$f(e^4) = \frac{e^4}{1-\log e^4} = \frac{e^4}{1-4\log e} = \frac{e^4}{1-4} = -\frac{e^4}{3}$$

The graph of the function is shown in Figure 11.29.

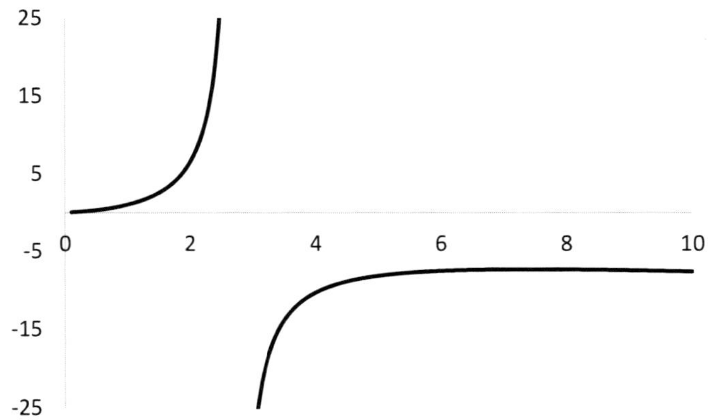

Figure 11.29. Graph of the function $f(x)=\dfrac{x}{1-\log x}$.

11.2.30. Problem 11.30

Study the function:

$$f(x)=\frac{1-\log(x-1)}{x-1}$$

11.2.30.1. Domain of the Function

The domain of the function is given by the intersection of the function domains in the numerator and the denominator, minus the values in which the denominator is equal to 0.

The domain of the numerator is determined by the domain of the logarithm function, whose argument must be:

$$x-1>0 \quad \Rightarrow \quad x>1$$

The domain of the logarithm function and consequently the domain of the numerator is:

$$D=\left]1,+\infty\right[$$

The domain of the denominator is $\boldsymbol{R}$ and the value where the denominator equals 0 it is:

$$x-1=0 \quad \Rightarrow \quad x=1$$

value already excluded from the domain of the numerator.

The function is therefore defined by:

$$f:\left]1,+\infty\right[\rightarrow \boldsymbol{R}$$

11.2.30.1. Intersection with the Axes

We verify whether the function intersects the x, y axes:

$$x=0 \quad \rightarrow \quad f(0)=?$$
$$y=f(x)=0 \quad \rightarrow \quad x=?$$

The point $x = 0$ does not belong to the domain of the function, so the function does not intersect the y-axis.

For $y = 0$:

$$y=f(x)=\frac{1-\log(x-1)}{x-1}=0$$

The only solution is that for which the numerator is equal to 0:

$$1-\log(x-1)=0 \Rightarrow \log(x-1)=1 \Rightarrow$$
$$e^{\log(x-1)}=e^{1} \Rightarrow x-1=e \Rightarrow x=1+e$$

The function intersects the x-axis at the point $x=1+e$.

11.2.30.2. Asymptotes

There is a vertical asymptote if:

$$\lim_{x\to x_0+}\frac{1-\log(x-1)}{x-1}=\pm\infty$$
$$\lim_{x\to x_0-}\frac{1-\log(x-1)}{x-1}=\pm\infty$$

Given the function definition domain, we can check the limit for $x_0^{+} = 1^{+}$.

$$\lim_{x\to 1+}\frac{1-\log(x-1)}{x-1}=\frac{-(-\infty)}{0^+}=+\infty$$

There is a vertical asymptote for $x \to 1^+$.

There is a horizontal asymptote if:

$$\lim_{x\to -\infty}\frac{-\log(x+1)}{x+1}=l_1\ ,\ l_1\in \boldsymbol{R}$$

$$\lim_{x\to +\infty}\frac{-\log(x+1)}{x+1}=l_2,\ l_2\in \boldsymbol{R}$$

Given the function definition domain, for horizontal asymptotes we check the limit only for $x \to +\infty$.

For the function it holds:

$$\lim_{x\to +\infty}\frac{1-\log(x-1)}{x-1}=\frac{-\infty}{+\infty}$$

It is an indeterminate form, to solve it we can apply De L'Hôpital's rule:

$$\lim_{x\to +\infty}\frac{1-\log(x-1)}{x-1}=\lim_{x\to +\infty}\frac{\frac{d}{dx}(1-\log(x-1))}{\frac{d}{dx}(x-1)}=$$

$$=\lim_{x\to +\infty}\frac{-\frac{1}{x-1}}{1}=\lim_{x\to +\infty}-\frac{1}{x-1}=-\frac{1}{+\infty}=0$$

For $x \to +\infty$ there is a horizontal asymptote at $y = 0$.

Since the function is continuous on the definition domain and given the existence of a vertical and a horizontal asymptote, there can be no oblique asymptotes

11.2.30.3. Maximum and/or Minimum Points

We verify if the function has maximum or minimum points, checking if the first derivative is equal to 0 at some point and if for these points the second derivative is > 0 or < 0.

$$f'(x_i)=0$$

$$f''(x_i)>0 \quad or \quad f''(x_i)<0$$

First derivative:

$$f'(x) = \frac{d}{dx}\left(\frac{1-\log(x-1)}{x-1}\right) = \frac{-\frac{1}{x-1}\cdot(x-1)-[1-\log(x-1)]\cdot 1}{(x-1)^2} =$$

$$= \frac{-1-1+\log(x-1)}{(x-1)^2} = \frac{-2+\log(x-1)}{(x-1)^2}$$

$$f'(x) = \frac{\log(x-1)-2}{(x-1)^2} = 0$$

The only solution is that for which the numerator is equal to 0:

$$\log(x-1)-2 = 0 \Rightarrow \log(x-1) = 2 \Rightarrow$$

$$e^{\log(x-1)} = e^2 \Rightarrow x-1 = e^2 \Rightarrow x = 1+e^2$$

The first derivative is equal to 0 at the point $x = 1+e^2$.

Second derivative:

$$f''(x) = \frac{d}{dx}\left(\frac{\log(x-1)-2}{(x-1)^2}\right) = \frac{\frac{1}{x-1}\cdot(x-1)^2 - [\log(x-1)-2]\cdot 2(x-1)}{(x-1)^4} =$$

$$= \frac{(x-1)-2(x-1)[\log(x-1)-2]}{(x-1)^4} = \frac{(x-1)[1-2\log(x+1)+4]}{(x-1)^4} =$$

$$= \frac{5-2\log(x-1)}{(x-1)^3}$$

The second derivative, at the point x where the first derivative is equal to 0, it is:

$$f''(1+e^2) = \frac{5-2\log(1+e^2-1)}{(1+e^2-1)^3} = \frac{5-2\log(e^2)}{e^{2\cdot 3}} =$$

$$= \frac{5-4\log e}{e^6} = \frac{1}{e^6} > 0$$

$x = 1+e^2$ is a point of minimum.

The value of the function at the minimum point is:

$$f(1+e^2)=\frac{1-\log\left(1+e^2-1\right)}{1+e^2-1}=\frac{1-\log\left(e^2\right)}{e^2}=$$

$$=\frac{1-2\log e}{e^2}=\frac{1-2}{e^2}=-\frac{1}{e^2}=-e^{-2}$$

11.2.30.4. Inflection Points

We verify whether the function has inflection points, checking whether the second derivative is equal to 0:

$$f''(x)=\frac{5-2\log(x-1)}{(x-1)^3}=0$$

The only solution is that for which the numerator is equal to 0. To solve the equation we apply the inverse function of the logarithm:

$$5-2\log(x-1)=0\Rightarrow\log(x-1)=\frac{5}{2}\Rightarrow$$

$$e^{\log(x-1)}=e^{\frac{5}{2}}\Rightarrow x-1=e^{\frac{5}{2}}\Rightarrow x=1+e^{\frac{5}{2}}$$

The second derivative is equal to 0 at the point $x=1+e^{\frac{5}{2}}$.

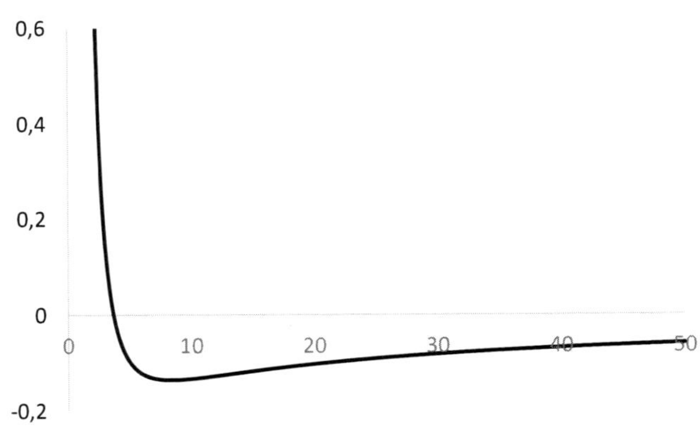

Figure 11.30. Graph of the function $f(x)=\frac{1-\log(x-1)}{x-1}$.

The value of the function at the inflection point is:

$$f(1+e^{\frac{5}{2}})=\frac{1-\log\left(1+e^{\frac{5}{2}}-1\right)}{1+e^{\frac{5}{2}}-1}=\frac{1-\log\left(e^{\frac{5}{2}}\right)}{e^{\frac{5}{2}}}=$$

$$=\frac{1-\frac{5}{2}\log e}{e^{\frac{5}{2}}}=\frac{1-\frac{5}{2}}{e^{\frac{5}{2}}}=\frac{-\frac{3}{2}}{e^{\frac{5}{2}}}-\frac{3}{2e^{\frac{5}{2}}}\equiv-\frac{3}{2}e^{-\frac{5}{2}}$$

The graph of the function is shown in Figure 11.30.

11.2.31. Problem 11.31

Study the function:

$$f(x)=\log(\sin x)$$

11.2.31.1. Domain of the Function

The domain of the function is determined by the values for which the argument of the logarithm function is > 0:

$$\sin x>0 \quad \Rightarrow$$
$$D=\{x\in \boldsymbol{R}:2k\pi<x<(2k+1)\pi,k\in I\}\equiv\left]2k\pi,(2k+1)\pi\right[,k\in I$$

The function is therefore defined by:

$$f:\left]2k\pi,(2k+1)\pi\right[\to \boldsymbol{R} \qquad ,k\in I$$

11.2.31.2. Intersection with the Axes

We verify whether the function intersects the x, y axes:

$$x=0 \quad \to \quad f(0)=?$$
$$y=f(x)=0 \quad \to \quad x=?$$

The point $x = 0$ does not belong to the domain of the function, so the function does not intersect the y-axis.

For $y = 0$:

$$y = f(x) = \log(\sin x) = 0$$

To solve the equation we apply the inverse function of the logarithm:

$$e^{\log(\sin x)} = e^0 \quad \Rightarrow \quad \sin x = 1$$

The solutions are represented by the set:

$$\left\{ \left(2k + \frac{1}{2} \right) \pi : k \in I \right\}$$

The elements of the set are the points where the function intersects the x-axis.

11.2.31.3. Asymptotes

There is a vertical asymptote if:

$$\lim_{x \to x_0 +} \log(\sin x) = \pm\infty$$
$$\lim_{x \to x_0 -} \log(\sin x) = \pm\infty$$

Given the function definition domain, we can check the limits for:

$$x \to 2k\pi^+ \text{ and } x \to (2k+1)\pi^-, \; k \in I$$

We obtain:

$$\lim_{x \to 2k\pi +} \log(\sin x) = -\infty$$
$$\lim_{x \to (2k+1)\pi -} \log(\sin x) = -\infty$$

There are vertical asymptotes at the periodic points:

$$2k\pi^+ \text{ and } (2k+1)\pi^-, \; k \in I$$

There is a horizontal asymptote if:

$$\lim_{x \to -\infty} \log(\sin x) = l_1 \ , \ l_1 \in \mathbf{R}$$

$$\lim_{x \to +\infty} \log(\sin x) = l_2 \ , \ l_2 \in \mathbf{R}$$

The argument of the logarithm function is a periodic and bounded function:

$$\sin x \in]0,1] \qquad \forall x \in]2k\pi, (2k+1)\pi[\qquad , k \in I$$

so there are no horizontal asymptotes.

Similar considerations on the set of infinite open and bounded intervals that constitute the domain of the function, allow us to conclude that there are no oblique asymptotes.

11.2.31.4. Maximum and/or Minimum Points

We verify if the function has maximum or minimum points, checking if the first derivative is equal to 0 at some point and if for these points the second derivative is > 0 or < 0.

$$f'(x_i) = 0$$

$$f''(x_i) > 0 \qquad or \qquad f''(x_i) < 0$$

First derivative:

$$f'(x) = \frac{d}{dx}\log(\sin x) = \frac{1}{\sin x} \cdot \cos x = \frac{\cos x}{\sin x}$$

$$f'(x) = \frac{\cos x}{\sin x} = 0 \qquad \Rightarrow \qquad \cos x = 0$$

The first derivative is equal to 0 at the points:

$$x_i = \left(2k + \frac{1}{2}\right)\pi \qquad , k \in I$$

It can be observed that the cosine would be equal to 0 even in the points represented by the set:

$$\left\{\left(2k-\frac{1}{2}\right)\pi : k \in I\right\}$$

However, these points must be excluded because they do not belong to the function definition domain.

Second derivative:

$$f''(x) = \frac{d}{dx}\left(\frac{\cos x}{\sin x}\right) = \frac{-\sin x \cdot \sin x - \cos x \cdot \cos x}{\sin^2 x} =$$

$$= -\frac{\sin^2 x + \cos^2 x}{\sin^2 x} = -\frac{1}{\sin^2 x}$$

The second derivative, at the points x where the first derivative is equal to 0, it is:

$$f''\left[\left(2k+\frac{1}{2}\right)\pi\right] = -\frac{1}{\sin^2\left[\left(2k+\frac{1}{2}\right)\pi\right]} = -1 < 0 \qquad ,k \in I$$

The function has periodic points of maximum at the values:

$$x_i = \left(2k+\frac{1}{2}\right)\pi \qquad ,k \in I$$

The values of the function at the maximum points are:

$$f\left[\left(2k+\frac{1}{2}\right)\pi\right] = \log\left[\sin\left((2k+\frac{1}{2})\pi\right)\right] = 0 \qquad ,k \in I$$

which coincide with the points of intersection with the x-axis as seen previously.

11.2.31.5. Inflection Points

We verify whether the function has inflection points, checking whether the second derivative is equal to 0:

$$f''(x) = -\frac{1}{\sin^2 x} = 0$$

The equation does not admit solutions, so there are no inflection points. The graph of the function is shown in Figure 11.31.

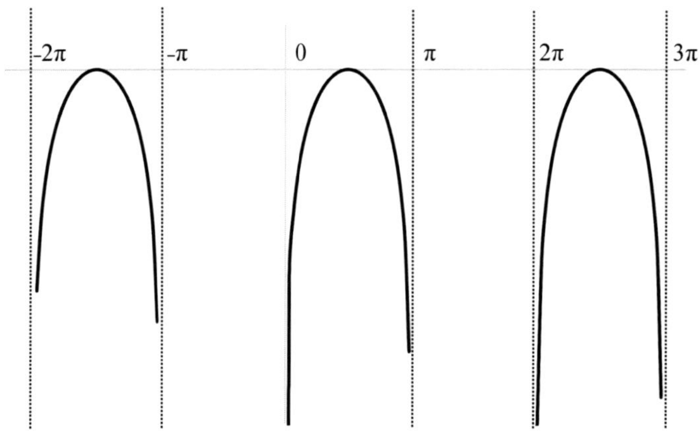

Figure 11.31. Graph of the function $f(x) = \log(\sin x)$.

Chapter 12

CALCULATION OF INTEGRALS

12.1. PROBLEMS

Calculate the following integrals.

$$\int_{0}^{2\pi} \sin x dx \tag{12.1}$$

$$\int_{0}^{\pi} \sin x dx \tag{12.2}$$

$$\int_{0}^{1} \frac{e^{-3x}}{4+2e^{-3x}} dx \tag{12.3}$$

$$\int_{-\pi}^{\pi} x \cdot \cos x dx \tag{12.4}$$

$$\int_{0}^{3} \frac{dx}{\sqrt[4]{4-x}} \tag{12.5}$$

$$\int_{0}^{1} \frac{e^{x}-e^{-x}}{e^{x}+e^{-x}} dx \tag{12.6}$$

$$\int_{-\frac{\pi}{2}}^{\frac{\pi}{2}} x \cdot \sin x dx \tag{12.7}$$

$$\int_{-1}^{1} \frac{e^{-5x}}{1+e^{-5x}} dx \tag{12.8}$$

$$\int_{1}^{e} x^3 \cdot \log x dx \tag{12.9}$$

$$\int_{0}^{\frac{\pi}{2}} \sqrt{1+\sin x} \cdot \cos x dx \tag{12.10}$$

$$\int_{-\frac{\pi}{2}}^{\frac{\pi}{2}} \sqrt{1+\sin x} \cdot \cos x dx \tag{12.11}$$

$$\int_{-\frac{\pi}{2}}^{\frac{\pi}{2}} \frac{\cos x}{\sqrt{1+\sin x}} dx \tag{12.12}$$

$$\int_{0}^{\pi} \frac{\sin x}{\sqrt{1+\cos x}} dx \tag{12.13}$$

$$\int_{\frac{\pi}{4}}^{\frac{3}{4}\pi} \log(\sin x) \cdot \cos x dx \tag{12.14}$$

$$\int_{0}^{\frac{\pi}{2}} \cos^n x \sin x dx \tag{12.15}$$

$$\int_{1}^{2} \frac{e^{x}+e^{-x}}{e^{x}-e^{-x}} dx \tag{12.16}$$

$$\int_{0}^{\frac{\pi}{2}} \sqrt[5]{\cos x - 1} \cdot \sin x dx \tag{12.17}$$

$$\int_{1}^{e} \frac{\log x}{x^{3}} dx \tag{12.18}$$

$$\int_{0}^{\pi} x^{3} \cdot \sin x dx \tag{12.19}$$

$$\int_{0}^{\frac{\pi}{4}} \sin^{3}(2x) dx \tag{12.20}$$

$$\int_{0}^{\frac{\pi}{2}} x(1+\cos x) dx \tag{12.21}$$

$$\int_{0}^{1} \frac{\log(x+1)}{x+1} dx \tag{12.22}$$

$$\int_{0}^{1} (x+1)^{2} \cdot \log(x+1) dx \tag{12.23}$$

$$\int_{-\frac{\pi}{2}}^{\frac{\pi}{2}} \cos(x+1) \cdot \sin^{3}(x+1) dx \tag{12.24}$$

$$\int_{0}^{\frac{\pi}{2}} \cos x \cdot e^{(\sin x+1)} dx \tag{12.25}$$

$$\int_0^{\frac{\pi}{2}} \sin x \cdot e^x dx \tag{12.26}$$

$$\int_1^4 x^{\frac{1}{2}} \cdot \log x dx \tag{12.27}$$

$$\int_0^1 x^2 \cdot e^x dx \tag{12.28}$$

$$\int_0^1 \log x dx \tag{12.29}$$

$$\int_1^{+\infty} \frac{1}{x^2} dx \tag{12.30}$$

$$\int_0^{\frac{1}{2}} \frac{dx}{\sqrt{1-4x^2}} \tag{12.31}$$

$$\int_0^{+\infty} e^{-\frac{x}{3}} dx \tag{12.32}$$

$$\int_{-\infty}^0 e^{2x} dx \tag{12.33}$$

$$\int_0^{+\infty} e^{-\left(\frac{x+1}{4}\right)} dx \tag{12.34}$$

$$\int_0^{+\infty} xe^{-x^2} dx \tag{12.35}$$

12.2. Solutions

12.2.1. Problem 12.1

Calculate the definite integral:

$$\int_0^{2\pi} \sin x dx$$

The domain of the integrand function is ***R***.

The integration interval $[0,2\pi]$ is within the domain of the integrand function on which the function is continuous, so the function is integrable on the interval.

If a function *f* admits primitive *F*, then it holds (Eq. 8.28):

$$\int_a^b f(x)dx = [F(x)]_a^b = F(b) - F(a)$$

In the case under consideration:

$$\int_0^{2\pi} \sin x dx = [-\cos x]_0^{2\pi} = -\cos 2\pi - (-\cos 0) =$$
$$= -\cos 2\pi + \cos 0 = -1 + 1 = 0$$

12.2.2. Problem 12.2

Calculate the definite integral:

$$\int_0^{\pi} \sin x dx$$

The domain of the integrand function is ***R***.

The integration interval $[0,\pi]$ is within the domain of the integrand function on which the function is continuous, so the function is integrable on the interval.

The primitive of the integrand function is the same as for problem 12.1. In this case it holds:

$$\int_0^{\pi} \sin x dx = [-\cos x]_0^{\pi} = -\cos\pi - (-\cos 0) =$$
$$= -\cos\pi + \cos 0 = -(-1) + 1 = 2$$

12.2.3. Problem 12.3

Calculate the definite integral:

$$\int_0^1 \frac{e^{-3x}}{4+2e^{-3x}} dx$$

The domain of the integrand function is ***R***.

The integration interval $[0,1]$ is within the domain of the integrand function on which the function is continuous, so the function is integrable on the interval.

The integral can be calculated by substitution (section 7.4):

$t = 4 + 2e^{-3x}$ (where $t > 0$)

from which:

$$dt = \frac{d}{dx}\left(4 + 2e^{-3x}\right)dx = -6e^{-3x}dx \quad \rightarrow \quad e^{-3x}dx = -\frac{1}{6}dt$$

The limits of integration are:

$$x = 0 \rightarrow t_{\inf} = 4 + 2e^0 = 4 + 2 = 6$$
$$x = 1 \rightarrow t_{\sup} = 4 + 2e^{-3}$$

The integral becomes:

$$\int_0^1 \frac{e^{-3x}}{4+2e^{-3x}} dx = \int_6^{4+2e^{-3}} -\frac{1}{6}\frac{dt}{t} = \frac{1}{6}\int_{4+2e^{-3}}^{6} \frac{dt}{t}$$

We have:

$$\frac{1}{6}\int_{4+2e^{-3}}^{6}\frac{dt}{t}=\frac{1}{6}\left[\log t\right]_{4+2e^{-3}}^{6}=\frac{1}{6}\left[\log 6-\log\left(4+2e^{-3}\right)\right]=$$

$$=\frac{1}{6}\left[\log 6+\log\left(\left(4+2e^{-3}\right)^{-1}\right)\right]=\frac{1}{6}\log\left(\frac{6}{4+2e^{-3}}\right)$$

The integral is therefore:

$$\int_{0}^{1}\frac{e^{-3x}}{4+2e^{-3x}}dx=\frac{1}{6}\log\left(\frac{6}{4+2e^{-3}}\right)$$

12.2.4. Problem 12.4

Calculate the definite integral:

$$\int_{-\pi}^{\pi}x\cdot\cos x dx$$

The domain of the integrand function is ***R***.

The integration interval $\left[-\pi,\pi\right]$ is within the domain of the integrand function on which the function is continuous, so the function is integrable on the interval.

The integral can be calculated by parts (section 7.5):

$$\int f(x)\cdot g(x)dx=f(x)\cdot G(x)-\int f'(x)\cdot G(x)dx$$

In the case under consideration we assume that:

$$f(x)=x \quad , \quad g(x)=\cos x$$
$$f'(x)=1 \quad , \quad G(x)=\sin x$$

We have:

$$\int_{-\pi}^{\pi} x \cdot \cos dx = [x \cdot \sin x]_{-\pi}^{\pi} - \int_{-\pi}^{\pi} \sin x dx = [x \cdot \sin x]_{-\pi}^{\pi} - [-\cos x]_{-\pi}^{\pi} =$$

$$= [x \cdot \sin x]_{-\pi}^{\pi} + [\cos x]_{-\pi}^{\pi} = \pi \cdot \sin \pi - (-\pi) \cdot \sin(-\pi) + \cos \pi - \cos(-\pi) =$$

$$= \pi \cdot 0 + \pi \cdot 0 + (-1) - (-1) = 0$$

The integral is therefore:

$$\int_{-\pi}^{\pi} x \cdot \cos x dx = 0$$

12.2.5. Problem 12.5

Calculate the definite integral:

$$\int_{0}^{3} \frac{dx}{\sqrt[4]{4-x}}$$

The domain of the integrand function, given the presence of a fourth root in the denominator, is determined by the values for which:

$$4 - x > 0 \quad \Rightarrow \quad x < 4$$

The function is therefore defined by:

$$f(x) = \frac{1}{\sqrt[4]{4-x}}, \; f:]-\infty, 4[\to \boldsymbol{R}$$

The integration interval $[0,3]$ is within the domain of the integrand function on which the function is continuous, so the function is integrable on the interval.

The integral can be calculated by substitution:

$$t = 4 - x$$

from which:

$$dt = \frac{d}{dx}(4-x)dx = -dx \qquad \rightarrow \qquad dx = -dt$$

The limits of integration are:

$$x = 0 \qquad \rightarrow \qquad t_{\inf} = 4 - 0 = 4$$
$$x = 3 \qquad \rightarrow \qquad t_{\sup} = 4 - 3 = 1$$

The integral becomes:

$$\int_0^3 \frac{dx}{\sqrt[4]{4-x}} = -\int_4^1 \frac{dt}{\sqrt[4]{t}} = \int_1^4 \frac{dt}{\sqrt[4]{t}}$$

We have:

$$\int_1^4 \frac{dt}{\sqrt[4]{t}} = \int_1^4 \frac{dt}{t^{\frac{1}{4}}} = \int_1^4 t^{-\frac{1}{4}} dt = \left[\frac{1}{-\frac{1}{4}+1} t^{-\frac{1}{4}+1}\right]_1^4$$

$$= \left[\frac{4}{3} t^{\frac{3}{4}}\right]_1^4 = \frac{4}{3}\left(4^{\frac{3}{4}} - 1^{\frac{3}{4}}\right) = \frac{4}{3}\left(2^{\frac{3}{2}} - 1\right)$$

The integral is therefore:

$$\int_0^3 \frac{dx}{\sqrt[4]{4-x}} = \frac{4}{3}\left(2^{\frac{3}{2}} - 1\right)$$

12.2.6. Problem 12.6

Calculate the definite integral:

$$\int_0^1 \frac{e^x - e^{-x}}{e^x + e^{-x}} dx$$

The domain of the integrand function is ***R***.

The integration interval $[0,1]$ is within the domain of the integrand function on which the function is continuous, so the function is integrable on the interval.

The integral can be calculated by substitution:

$t = e^x + e^{-x}$

from which:

$$dt = \frac{d}{dx}\left(e^x + e^{-x}\right)dx = \left(e^x - e^{-x}\right)dx$$

The limits of integration are:

$$x = 0 \quad \rightarrow \quad t_{\text{inf}} = e^0 + e^0 = 2$$

$$x = 1 \quad \rightarrow \quad t_{\text{sup}} = e + e^{-1}$$

The integral becomes:

$$\int_0^1 \frac{e^x - e^{-x}}{e^x + e^{-x}}dx = \int_2^{e+e^{-1}} \frac{dt}{t}$$

We have:

$$\int_2^{e+e^{-1}} \frac{dt}{t} = [\log t]_2^{e+e^{-1}} = \log\left(e + e^{-1}\right) - \log 2 = \log\left(\frac{e + e^{-1}}{2}\right)$$

The integral is therefore:

$$\int_0^1 \frac{e^x - e^{-x}}{e^x + e^{-x}}dx = \log\left(\frac{e + e^{-1}}{2}\right)$$

12.2.7. Problem 12.7

Calculate the definite integral:

$$\int_{-\frac{\pi}{2}}^{\frac{\pi}{2}} x \cdot \sin x dx$$

The domain of the integrand function is ***R***.

The integration interval $\left[-\frac{\pi}{2}, \frac{\pi}{2}\right]$ is within the domain of the integrand function on which the function is continuous, so the function is integrable on the interval.

The integral can be calculated by parts:

$$\int f(x) \cdot g(x) dx = f(x) \cdot G(x) - \int f'(x) \cdot G(x) dx$$

We assume that:

$$f(x) = x \quad , \quad g(x) = \sin x$$
$$f'(x) = 1 \quad , \quad G(x) = -\cos x$$

We have:

$$\int_{-\frac{\pi}{2}}^{\frac{\pi}{2}} x \cdot \sin dx = \left[x \cdot (-\cos x)\right]_{-\frac{\pi}{2}}^{\frac{\pi}{2}} - \int_{-\frac{\pi}{2}}^{\frac{\pi}{2}} -\cos x dx =$$

$$= \left[-x \cdot \cos x\right]_{-\frac{\pi}{2}}^{\frac{\pi}{2}} + \int_{-\frac{\pi}{2}}^{\frac{\pi}{2}} \cos x dx = -\left[x \cdot \cos x\right]_{-\frac{\pi}{2}}^{\frac{\pi}{2}} + \left[\sin x\right]_{-\frac{\pi}{2}}^{\frac{\pi}{2}} =$$

$$= -\left[\frac{\pi}{2} \cdot \cos \frac{\pi}{2} - \left(-\frac{\pi}{2}\right) \cdot \cos\left(-\frac{\pi}{2}\right)\right] + \sin \frac{\pi}{2} - \sin\left(-\frac{\pi}{2}\right) =$$

$$= -\left[\frac{\pi}{2} \cdot 0 - \left(-\frac{\pi}{2}\right) \cdot 0\right] + 1 - (-1) = 1 + 1 = 2$$

The integral is therefore:

$$\int_{-\frac{\pi}{2}}^{\frac{\pi}{2}} x \cdot \sin x dx = 2$$

12.2.8. Problem 12.8

Calculate the definite integral:

$$\int_{-1}^{1} \frac{e^{-5x}}{1+e^{-5x}} dx$$

The domain of the integrand function is ***R***.

The integration interval $[-1,1]$ is within the domain of the integrand function on which the function is continuous, so the function is integrable on the interval.

The integral can be calculated by substitution:

$$t = 1 + e^{-5x} \quad \Rightarrow \quad dt = -5e^{-5x} dx \quad \Rightarrow \quad e^{-5x} dx = -\frac{1}{5} dt$$

The limits of integration are:

$$x = 1 \quad \rightarrow \quad t_{\text{sup}} = 1 + e^{-5}$$

$$x = -1 \quad \rightarrow \quad t_{\text{inf}} = 1 + e^{5}$$

We have:

$$\int_{-1}^{1} \frac{e^{-5x}}{1+e^{-5x}} dx = \int_{1+e^{5}}^{1+e^{-5}} -\frac{1}{5}\frac{dt}{t} = -\frac{1}{5}\int_{1+e^{5}}^{1+e^{-5}} \frac{dt}{t} = \frac{1}{5}\int_{1+e^{-5}}^{1+e^{5}} \frac{dt}{t} = \frac{1}{5}[\log t]_{1+e^{-5}}^{1+e^{5}} =$$

$$= \frac{1}{5}\left[\log(1+e^{5}) - \log(1+e^{-5})\right] = \frac{1}{5}\log\left(\frac{1+e^{5}}{1+e^{-5}}\right) = \frac{1}{5}\log\left(\frac{1+e^{5}}{e^{-5}(e^{5}+1)}\right) =$$

$$= \frac{1}{5}\log\left(\frac{1}{e^{-5}}\right) = \frac{1}{5}\log(e^{5}) = \frac{1}{5}\cdot 5\log e = 1$$

The integral is therefore:

$$\int_{-1}^{1} \frac{e^{-5x}}{1+e^{-5x}} dx = 1$$

12.2.9. Problem 12.9

Calculate the definite integral:

$$\int_1^e x^3 \cdot \log x dx$$

The domain of the integrand function is $\boldsymbol{R}^+$.

The integration interval $[1, e]$ is within the domain of the integrand function on which the function is continuous, so the function is integrable on the interval.

The integral can be calculated by parts:

$$\int f(x) \cdot g(x) dx = f(x) \cdot G(x) - \int f'(x) \cdot G(x) dx$$

We assume that:

$$f(x) = \log x \quad , \quad g(x) = x^3$$
$$f'(x) = \frac{1}{x} \quad , \quad G(x) = \frac{1}{4} x^4$$

We have:

$$\int_1^e x^3 \cdot \log x dx = \left[\log x \cdot \frac{x^4}{4} \right]_1^e - \int_1^e \frac{1}{x} \cdot \frac{x^4}{4} dx = \left[\log x \cdot \frac{x^4}{4} \right]_1^e - \int_1^e \frac{x^3}{4} dx =$$
$$= \left[\log x \cdot \frac{x^4}{4} \right]_1^e - \left[\frac{1}{4} \cdot \frac{x^4}{4} \right]_1^e = \log e \cdot \frac{e^4}{4} - \log 1 \cdot \frac{1^4}{4} - \left(\frac{e^4}{16} - \frac{1^4}{16} \right) =$$
$$= 1 \cdot \frac{e^4}{4} - 0 \cdot \frac{1}{4} - \frac{e^4}{16} + \frac{1}{16} = \frac{e^4}{4} - \frac{e^4}{16} + \frac{1}{16} = e^4 \left(\frac{1}{4} - \frac{1}{16} \right) + \frac{1}{16} =$$
$$= \frac{3e^4}{16} + \frac{1}{16} = \frac{1}{16} \left(3e^4 + 1 \right)$$

The integral is therefore:

$$\int_1^e x^3 \cdot \log x dx = \frac{1}{16} \left(3e^4 + 1 \right)$$

12.2.10. Problem 12.10

Calculate the definite integral:

$$\int_0^{\frac{\pi}{2}} \sqrt{1+\sin x} \cdot \cos x dx$$

The domain of the integrand function is given by the values for which the argument of the square root function is ≥ 0. In the specific case by the values:

$$x \in \boldsymbol{R}: \quad \sin x \geq -1$$

Since $|\sin x| \leq 1$, the domain is $\boldsymbol{R}$.

The integration interval $\left[0, \frac{\pi}{2}\right]$ is within the domain of the integrand function on which the function is continuous, so the function is integrable on the interval.

The integral can be calculated by substitution:

$$t = 1 + \sin x$$

$$dt = \frac{d}{dx}(1+\sin x)dx = \cos x dx$$

The limits of integration are:

$$x = \frac{\pi}{2} \quad \rightarrow \quad t_{sup} = 1 + \sin\frac{\pi}{2} = 2$$

$$x = 0 \quad \rightarrow \quad t_{inf} = 1 + \sin 0 = 1$$

The integral becomes:

$$\int_0^{\frac{\pi}{2}} \sqrt{1+\sin x} \cdot \cos x dx = \int_1^2 \sqrt{t} dt$$

$$\int_1^2 \sqrt{t}dt = \int_1^2 t^{\frac{1}{2}}dt = \left[\frac{1}{\frac{1}{2}+1}t^{\frac{1}{2}+1}\right]_1^2 = \left[\frac{2}{3}t^{\frac{3}{2}}\right]_1^2 =$$

$$= \frac{2}{3}\cdot\left(2^{\frac{3}{2}} - 1^{\frac{3}{2}}\right) = \frac{2}{3}\cdot\left(2\sqrt{2}-1\right)$$

The integral is therefore:

$$\int_0^{\frac{\pi}{2}} \sqrt{1+\sin x}\cdot\cos x dx = \frac{2}{3}\left(2\sqrt{2}-1\right)$$

12.2.11. Problem 12.11

Calculate the definite integral:

$$\int_{-\frac{\pi}{2}}^{\frac{\pi}{2}} \sqrt{1+\sin x}\cdot\cos x dx$$

The domain of the integrand function is given by the values for which the argument of the square root function is ≥ 0. In the specific case by the values:

$$x \in \boldsymbol{R}: \quad \sin x \geq -1$$

Since $|\sin x| \leq 1$, the domain is $\boldsymbol{R}$.

The integration interval $\left[-\frac{\pi}{2}, \frac{\pi}{2}\right]$ is within the domain of the integrand function on which the function is continuous, so the function is integrable on the interval.

The integral can be calculated by substitution:

$$t = 1 + \sin x$$

$$dt = \frac{d}{dx}(1 + \sin x)dx = \cos x dx$$

The limits of integration are:

$$x = \frac{\pi}{2} \quad \rightarrow \quad t_{\text{sup}} = 1 + \sin\frac{\pi}{2} = 2$$

$$x = -\frac{\pi}{2} \quad \rightarrow \quad t_{\text{inf}} = 1 + \sin\left(-\frac{\pi}{2}\right) = 1 - 1 = 0$$

The integral becomes:

$$\int_{-\frac{\pi}{2}}^{\frac{\pi}{2}} \sqrt{1 + \sin x} \cdot \cos x dx = \int_0^2 \sqrt{t} dt$$

$$\int_0^2 \sqrt{t} dt = \int_0^2 t^{\frac{1}{2}} dt = \left[\frac{2}{3} t^{\frac{3}{2}}\right]_0^2 = \frac{2}{3} \cdot 2^{\frac{3}{2}} = \frac{2}{3} \cdot 2\sqrt{2} = \frac{4\sqrt{2}}{3}$$

The integral is therefore:

$$\int_{-\frac{\pi}{2}}^{\frac{\pi}{2}} \sqrt{1 + \sin x} \cdot \cos x dx = \frac{4\sqrt{2}}{3}$$

12.2.12. Problem 12.12

Calculate the definite integral:

$$\int_{-\frac{\pi}{2}}^{\frac{\pi}{2}} \frac{\cos x}{\sqrt{1 + \sin x}} dx$$

The domain of the integrand function is given by the values for which the argument of the square root function being in the denominator is > 0. In the specific case by the values:

$x \in \boldsymbol{R}: \quad \sin x > -1$

The domain is therefore:

$$\boldsymbol{R} - \left\{2k\pi - \frac{\pi}{2} : k \in I\right\} \equiv \boldsymbol{R} - \left\{\left(2k - \frac{1}{2}\right)\pi : k \in I\right\}$$

The lower bound of the interval of integration $\left[-\frac{\pi}{2}, \frac{\pi}{2}\right]$ is outside the domain of the integrand function, so the function is not integrable on the interval.

The integral can be calculated as (Eq. 8.32):

$$\int_{-\frac{\pi}{2}}^{\frac{\pi}{2}} \frac{\cos x}{\sqrt{1+\sin x}} dx = \lim_{\alpha \to -\frac{\pi}{2}^+} \int_{\alpha}^{\frac{\pi}{2}} \frac{\cos x}{\sqrt{1+\sin x}} dx$$

The integral can be calculated by substitution:

$$t = 1 + \sin x$$
$$dt = \frac{d}{dx}(1 + \sin x)dx = \cos x dx$$

The limits of integration are:

$$x = \frac{\pi}{2} \quad \to \quad t_{\sup} = 1 + \sin\frac{\pi}{2} = 2$$
$$x = \alpha \quad \to \quad t_{\inf} = 1 + \sin\alpha$$

The definite integral is given by:

$$\int_{-\frac{\pi}{2}}^{\frac{\pi}{2}} \frac{\cos x}{\sqrt{1+\sin x}} dx = \lim_{\alpha \to -\frac{\pi}{2}^+} \int_{\alpha}^{\frac{\pi}{2}} \frac{\cos x}{\sqrt{1+\sin x}} dx = \lim_{\alpha \to -\frac{\pi}{2}^+} \int_{1+\sin\alpha}^{2} \frac{dt}{\sqrt{t}}$$

$$\lim_{\alpha \to -\frac{\pi}{2}^+} \int_{1+\sin\alpha}^{2} \frac{dt}{\sqrt{t}} = \lim_{\alpha \to -\frac{\pi}{2}^+} \int_{1+\sin\alpha}^{2} t^{-\frac{1}{2}} dt = \lim_{\alpha \to -\frac{\pi}{2}^+} \left[\frac{1}{-\frac{1}{2}+1} t^{-\frac{1}{2}+1} \right]_{1+\sin\alpha}^{2} =$$

$$= \lim_{\alpha \to -\frac{\pi}{2}^+} \left[\frac{1}{\frac{1}{2}} t^{\frac{1}{2}} \right]_{1+\sin\alpha}^{2} =$$

$$= \lim_{\alpha \to -\frac{\pi}{2}^+} \left[2t^{\frac{1}{2}} \right]_{1+\sin\alpha}^{2} = \lim_{\alpha \to -\frac{\pi}{2}^+} \left[2 \cdot 2^{\frac{1}{2}} - 2(1+\sin\alpha)^{\frac{1}{2}} \right] =$$

$$= 2\sqrt{2} - \lim_{\alpha \to -\frac{\pi}{2}^+} 2(1+\sin\alpha)^{\frac{1}{2}} = 2\sqrt{2} - 0 = 2\sqrt{2}$$

The integral is therefore:

$$\int_{-\frac{\pi}{2}}^{\frac{\pi}{2}} \frac{\cos x}{\sqrt{1+\sin x}} dx = 2\sqrt{2}$$

12.2.13. Problem 12.13

Calculate the definite integral:

$$\int_{0}^{\pi} \frac{\sin x}{\sqrt{1+\cos x}} dx$$

The domain of the integrand function is given by the values for which the argument of the square root function being in the denominator is > 0. In the specific case by the values:

$$x \in \boldsymbol{R}: \quad \cos x > -1$$

The domain is therefore:

$$\boldsymbol{R} - \{2k\pi - \pi : k \in I\} \equiv \boldsymbol{R} - \{(2k-1)\pi : k \in I\}$$

The upper bound of the interval of integration $[0, \pi]$ is outside the domain of the integrand function, so the function is not integrable on the interval.

The integral can be calculated as (Eq. 8.35):

$$\int_0^{\pi} \frac{\sin x}{\sqrt{1+\cos x}} dx = \lim_{\beta \to \pi-} \int_0^{\beta} \frac{\sin x}{\sqrt{1+\cos x}} dx$$

The integral can be calculated by substitution:

$$t = 1 + \cos x$$

$$dt = \frac{d}{dx}(1+\cos x)dx = -\sin x dx \qquad \Rightarrow \qquad -dt = \sin x dx$$

The limits of integration are:

$$x = \beta \quad \to \quad t_{\sup} = 1 + \cos\beta$$

$$x = 0 \quad \to \quad t_{\inf} = 1 + \cos 0 = 1 + 1 = 2$$

The definite integral is given by:

$$\int_0^{\pi} \frac{\sin x}{\sqrt{1+\cos x}} dx = \lim_{\beta \to \pi-} \int_0^{\beta} \frac{\sin x}{\sqrt{1+\cos x}} dx =$$

$$= \lim_{\beta \to \pi-} \int_2^{1+\cos\beta} \frac{-dt}{\sqrt{t}} = \lim_{\beta \to \pi-} \int_{1+\cos\beta}^{2} \frac{dt}{\sqrt{t}}$$

We have (see problem 12.12):

$$\lim_{\beta\to\pi-}\int_{1+\cos\beta}^{2}\frac{dt}{\sqrt{t}}=\lim_{\beta\to\pi-}\int_{1+\cos\beta}^{2}t^{-\frac{1}{2}}dt=\lim_{\beta\to\pi-}\left[2t^{\frac{1}{2}}\right]_{1+\cos\beta}^{2}=$$

$$=\lim_{\beta\to\pi-}\left[2\cdot 2^{\frac{1}{2}}-2(1+\cos\beta)^{\frac{1}{2}}\right]=2\sqrt{2}-\lim_{\beta\to\pi-}2(1+\cos\beta)^{\frac{1}{2}}=$$

$$=2\sqrt{2}-0=2\sqrt{2}$$

The integral is therefore:

$$\int_{0}^{\pi}\frac{\sin x}{\sqrt{1+\cos x}}dx=2\sqrt{2}$$

12.2.14. Problem 12.14

Calculate the definite integral:

$$\int_{\frac{\pi}{4}}^{\frac{3}{4}\pi}\log(\sin x)\cdot\cos x dx$$

The domain of the integrand function is given by the values for which the argument of the logarithm function is > 0. In the specific case by the values:

$$x\in\boldsymbol{R}: \quad \sin x>0$$

The domain is therefore:

$$\{x\in\boldsymbol{R}:2k\pi<x<(2k+1)\pi,k\in I\}\equiv\,]2k\pi,(2k+1)\pi[,k\in I$$

The integration interval $\left[\frac{\pi}{4},\frac{3}{4}\pi\right]$ is within the domain of the integrand function on which the function is continuous, so the function is integrable on the interval.

The integral can be calculated by substitution:

$$t=\sin x \quad\Rightarrow\quad dt=\cos x dx$$

The limits of integration are:

$$x = \frac{3\pi}{4} \quad \rightarrow \quad t_{\sup} = \sin\frac{3\pi}{4}$$

$$x = \frac{\pi}{4} \quad \rightarrow \quad t_{\inf} = \sin\frac{\pi}{4}$$

The definite integral is given by:

$$\int_{\frac{\pi}{4}}^{\frac{3}{4}\pi} \log(\sin x)\cdot \cos x dx = \int_{\sin\frac{\pi}{4}}^{\sin\frac{3\pi}{4}} \log t dt$$

The integral thus obtained can be calculated by parts, as was done previously for the indefinite integral (Eq. 7.79, sub-subsection 7.5.1.3):

$$\int \log x dx = x(\log x - 1) + c$$

In this case:

$$\int_{\sin\frac{\pi}{4}}^{\sin\frac{3\pi}{4}} \log t dt = \left[t(\log t - 1)\right]_{\sin\frac{\pi}{4}}^{\sin\frac{3\pi}{4}} =$$

$$= \sin\frac{3\pi}{4}\left(\log\left(\sin\frac{3\pi}{4}\right) - 1\right) - \sin\frac{\pi}{4}\left(\log\left(\sin\frac{\pi}{4}\right) - 1\right)$$

Since:

$$\sin\frac{3\pi}{4} = \sin\frac{\pi}{4} = \frac{1}{\sqrt{2}}$$

it results:

$$\sin\frac{3\pi}{4}\left(\log\left(\sin\frac{3\pi}{4}\right) - 1\right) - \sin\frac{\pi}{4}\left(\log\left(\sin\frac{\pi}{4}\right) - 1\right) = 0$$

The integral is therefore:

$$\int_{\frac{\pi}{4}}^{\frac{3}{4}\pi} \log(\sin x)\cdot \cos x dx = \int_{\sin\frac{\pi}{4}}^{\sin\frac{3\pi}{4}} \log t dt = 0$$

Since the two limits of integration are the same, it can also be observed that the integral is zero without having to explicitly calculate it:

$$\int_{\sin\frac{\pi}{4}}^{\sin\frac{3\pi}{4}} \log t dt = 0$$

12.2.15. Problem 12.15

Calculate the definite integral:

$$\int_{0}^{\frac{\pi}{2}} \cos^n x \sin x dx \qquad , n \in N$$

In the case under examination, the integrand contains the cosine function as the argument of a power function with exponent > 0 (Eq. 2.6), so the following cases can be excluded:

$$\cos^{-\alpha} x = \frac{1}{\cos^{\alpha} x} \quad , \quad \alpha > 0$$

cases that could give rise to values for which the integrand function diverges, from which the need in these cases to verify the domain of the integrand with respect to the integration interval.

In this case, the domain of the integrand function is ***R***.

The integration interval $\left[0, \frac{\pi}{2}\right]$ is within the domain of the integrand function on which the function is continuous, so the function is integrable on the interval.

The integral can be calculated by substitution:

$$t = \cos x$$

$$dt = \frac{d}{dx}\cos x dx = -\sin x dx \qquad \Rightarrow \qquad -dt = \sin x dx$$

The limits of integration are:

$$x = \frac{\pi}{2} \qquad \rightarrow \qquad t_{\sup} = \cos\frac{\pi}{2} = 0$$

$$x = 0 \qquad \rightarrow \qquad t_{\inf} = \cos 0 = 1$$

The definite integral is given by:

$$\int_0^{\frac{\pi}{2}} \cos^n x \sin x dx = -\int_1^0 t^n dt = \int_0^1 t^n dt$$

$$\int_0^1 t^n dt = \left[\frac{1}{n+1}t^{n+1}\right]_0^1 = \frac{1}{n+1}\cdot\left(1^{n+1} - 0^{n+1}\right) = \frac{1}{n+1}$$

The integral is therefore:

$$\int_0^{\frac{\pi}{2}} \cos^n x \sin x dx = \frac{1}{n+1}$$

12.2.16. Problem 12.16

Calculate the definite integral:

$$\int_1^2 \frac{e^x + e^{-x}}{e^x - e^{-x}} dx$$

The domain of the integrand function is ***R*** minus the values in which the denominator is equal to zero:

$$e^x - e^{-x} = 0 \qquad \Rightarrow \qquad x = 0 \qquad \Rightarrow \qquad e^0 - e^0 = 1 - 1 = 0$$

The domain is therefore:

$$D = \boldsymbol{R} - \{0\}$$

The integration interval $[1,2]$ is within the domain of the integrand function on which the function is continuous, so the function is integrable on the interval.

The integral can be calculated by substitution:

$$t = e^x - e^{-x} \quad \Rightarrow$$

$$dt = \frac{d}{dx}\left(e^x - e^{-x}\right)dx = \left(e^x - \left(-e^{-x}\right)\right)dx = \left(e^x + e^{-x}\right)dx$$

The limits of integration are:

$$x = 1 \quad \rightarrow \quad t_{\text{inf}} = e^1 - e^{-1}$$

$$x = 2 \quad \rightarrow \quad t_{\text{sup}} = e^2 - e^{-2}$$

The integral becomes:

$$\int_1^2 \frac{e^x + e^{-x}}{e^x - e^{-x}}\,dx \quad = \quad \int_{e-e^{-1}}^{e^2-e^{-2}} \frac{dt}{t}$$

Since $t > 0$ in the integration interval, the integral is given by:

$$\int_{e-e^{-1}}^{e^2-e^{-2}} \frac{dt}{t} = [\log t]_{e-e^{-1}}^{e^2-e^{-2}} = \log\left(e^2 - e^{-2}\right) - \log\left(e - e^{-1}\right) =$$

$$= \log\left(\frac{e^2 - e^{-2}}{e - e^{-1}}\right) = \log\left(\frac{\left(e + e^{-1}\right)\cdot\left(e - e^{-1}\right)}{e - e^{-1}}\right) = \log\left(e + e^{-1}\right)$$

The integral is therefore:

$$\int_1^2 \frac{e^x + e^{-x}}{e^x - e^{-x}}\,dx = \log\left(e + e^{-1}\right)$$

12.2.17. Problem 12.17

Calculate the definite integral:

$$\int_0^{\frac{\pi}{2}} \sqrt[5]{\cos x - 1} \cdot \sin x dx$$

The domain of the integrand function is ***R***.

The integration interval $\left[0, \frac{\pi}{2}\right]$ is within the domain of the integrand function on which the function is continuous, so the function is integrable on the interval.

The integral can be calculated by substitution:

$$t = \cos x - 1 \quad \Rightarrow$$

$$dt = \frac{dt}{dx} dx = \frac{d}{dx}(\cos x - 1)dx = -\sin x dx$$

The limits of integration are:

$$x = 0 \quad \rightarrow \quad t_{\inf} = \cos 0 - 1 = 1 - 1 = 0$$

$$x = \frac{\pi}{2} \quad \rightarrow \quad t_{\sup} = \cos\frac{\pi}{2} - 1 = 0 - 1 = -1$$

The definite integral is given by:

$$\int_0^{\frac{\pi}{2}} \sqrt[5]{\cos x - 1} \cdot \sin x dx = \int_0^{\frac{\pi}{2}} (\cos x - 1)^{\frac{1}{5}} \cdot \sin x dx = \int_0^{-1} -t^{\frac{1}{5}} dt = \int_{-1}^{0} t^{\frac{1}{5}} dt =$$

$$= \left[\frac{1}{\frac{1}{5}+1} t^{\frac{1}{5}+1}\right]_{-1}^{0} = \left[\frac{5}{6} t^{\frac{6}{5}}\right]_{-1}^{0} = \frac{5}{6}\left[0^{\frac{6}{5}} - (-1)^{\frac{6}{5}}\right] = -\frac{5}{6}$$

The integral is:

$$\int_0^{\frac{\pi}{2}} \sqrt[5]{\cos x - 1} \cdot \sin x dx = -\frac{5}{6}$$

12.2.18. Problem 12.18

Calculate the definite integral:

$$\int_1^e \frac{\log x}{x^3} dx$$

The domain of the integrand function is $\boldsymbol{R}^+$.

The integration interval $[1, e]$ is within the domain of the integrand function on which the function is continuous, so the function is integrable on the interval.

The integral can be calculated by parts:

$$\int f(x) \cdot g(x) dx = f(x) \cdot G(x) - \int f'(x) \cdot G(x) dx$$

We assume that:

$$f(x) = \log x \quad , \quad g(x) = x^{-3}$$
$$f'(x) = \frac{1}{x} \quad , \quad G(x) = \frac{1}{-3+1} x^{-3+1} = -\frac{1}{2} x^{-2}$$

The definite integral is given by:

$$\int_1^e x^{-3} \cdot \log x dx = \left[\log x \cdot \left(-\frac{x^{-2}}{2} \right) \right]_1^e - \int_1^e \frac{1}{x} \cdot \left(-\frac{x^{-2}}{2} \right) dx =$$
$$= \left[-\frac{1}{2} x^{-2} \cdot \log x \right]_1^e + \frac{1}{2} \int_1^e x^{-3} dx = \left[-\frac{1}{2} x^{-2} \cdot \log x \right]_1^e + \frac{1}{2} \left[-\frac{1}{2} \cdot x^{-2} \right]_1^e =$$
$$= -\frac{1}{2} e^{-2} \log e + \frac{1}{2} \cdot 1^{-2} \log 1 - \frac{1}{4} e^{-2} + \frac{1}{4} 1^{-2} =$$
$$= -\frac{1}{2} e^{-2} - \frac{1}{4} e^{-2} + \frac{1}{4} = -\frac{3}{4} e^{-2} + \frac{1}{4} = \frac{1}{4} \left(1 - 3e^{-2} \right)$$

The integral is therefore:

$$\int_1^e \frac{\log x}{x^3}dx = \int_1^e x^{-3} \cdot \log x dx = \frac{1}{4}\left(1-3e^{-2}\right)$$

12.2.19. Problem 12.19

Calculate the definite integral:

$$\int_0^\pi x^3 \cdot \sin x dx$$

The domain of the integrand function is ***R***.

The integration interval $[0, \pi]$ is within the domain of the integrand function on which the function is continuous, so the function is integrable on the interval.

The integral can be calculated by parts:

$$\int f(x) \cdot g(x)dx = f(x) \cdot G(x) - \int f'(x) \cdot G(x)dx$$

We assume that:

$$f(x) = x^3 \quad , \quad g(x) = \sin x$$
$$f'(x) = 3x^2 \quad , \quad G(x) = -\cos x$$

The definite integral is given by:

$$\int_0^\pi x^3 \cdot \sin x dx = \left[x^3 \cdot (-\cos x)\right]_0^\pi - \int_0^\pi 3x^2 \cdot (-\cos x)dx =$$

$$= -\left[x^3 \cdot \cos x\right]_0^\pi + \int_0^\pi 3x^2 \cdot \cos x dx$$

where, by iterating the integration by parts:

$$\int_0^\pi 3x^2 \cdot \cos x dx = \left[3x^2 \cdot \sin x\right]_0^\pi - \int_0^\pi 6x \cdot \sin x dx$$

and:

$$-\int_0^{\pi} 6x\cdot\sin x dx = -\left\{[6x\cdot(-\cos x)]_0^{\pi} - \int_0^{\pi} 6\cdot(-\cos x)dx\right\} =$$

$$= [6x\cdot\cos x]_0^{\pi} - \int_0^{\pi} 6\cos x dx = [6x\cdot\cos x]_0^{\pi} - [6\sin x]_0^{\pi}$$

we obtain thus:

$$\int_0^{\pi} x^3\cdot\sin x dx = -[x^3\cdot\cos x]_0^{\pi} + [3x^2\cdot\sin x]_0^{\pi} + [6x\cdot\cos x]_0^{\pi} - [6\sin x]_0^{\pi} =$$
$$= -[\pi^3\cdot\cos\pi - 0^3\cdot\cos 0] + [3\pi^2\cdot\sin\pi - 3\cdot 0^2\cdot\sin 0] +$$
$$+ [6\pi\cdot\cos\pi - 6\cdot 0\cdot\cos 0] - [6\sin\pi - 6\sin 0] = \pi^3 - 6\pi$$

The integral is therefore:

$$\int_0^{\pi} x^3\cdot\sin x dx = \pi^3 - 6\pi$$

12.2.20. Problem 12.20

Calculate the definite integral:

$$\int_0^{\frac{\pi}{4}} \sin^3(2x)dx$$

The domain of the integrand function is ***R***.

The integration interval $\left[0,\frac{\pi}{4}\right]$ is within the domain of the integrand function on which the function is continuous, so the function is integrable on the interval.

The integral can be written as:

$$\int_0^{\frac{\pi}{4}} \sin^3(2x)dx = \int_0^{\frac{\pi}{4}} \sin^2(2x)\sin(2x)dx$$

Considering that:

$$\sin^2(2x)+\cos^2(2x)=1 \quad \Rightarrow \quad \sin^2(2x)=1-\cos^2(2x)$$

we obtain:

$$\int_0^{\frac{\pi}{4}} \sin^2(2x)\sin(2x)dx = \int_0^{\frac{\pi}{4}} \left(1-\cos^2(2x)\right)\cdot \sin(2x)dx =$$

$$= \int_0^{\frac{\pi}{4}} \sin(2x)dx - \int_0^{\frac{\pi}{4}} \cos^2(2x)\sin(2x)dx$$

For the first integral of the sum, a primitive of the function:

$$f(x) = F'(x) = \sin(2x)$$

it is:

$$F(x) = -\frac{1}{2}\cos(2x)$$

Thus:

$$\int_0^{\frac{\pi}{4}} \sin(2x)dx = \left[-\frac{1}{2}\cos(2x)\right]_0^{\frac{\pi}{4}} = -\frac{1}{2}\left[\cos\left(2\cdot\frac{\pi}{4}\right) - \cos(2\cdot 0)\right] =$$

$$= -\frac{1}{2}\left[\cos\frac{\pi}{2} - \cos 0\right] = -\frac{1}{2}(0-1) = \frac{1}{2}$$

For the second integral of the sum, the integral can be calculated by substitution:

$$t = \cos(2x) \quad \Rightarrow \quad dt = -2\sin(2x)dx \quad \Rightarrow \quad -\frac{1}{2}dt = \sin(2x)dx$$

The limits of integration are:

$$x = 0 \quad \rightarrow \quad t_{\inf} = \cos(2 \cdot 0) = \cos 0 = 1$$

$$x = \frac{\pi}{4} \quad \rightarrow \quad t_{\sup} = \cos\left(2 \cdot \frac{\pi}{4}\right) = \cos\frac{\pi}{2} = 0$$

The definite integral is given by:

$$-\int_0^{\frac{\pi}{4}} \cos^2(2x)\sin(2x)dx = -\int_1^0 -\frac{1}{2}t^2 dt = \frac{1}{2}\int_1^0 t^2 dt = -\frac{1}{2}\int_0^1 t^2 dt =$$

$$= -\frac{1}{2}\left[\frac{1}{3}t^3\right]_0^1 = -\frac{1}{6}\left[t^3\right]_0^1 = -\frac{1}{6}$$

We obtain therefore:

$$\int_0^{\frac{\pi}{4}} \sin^3(2x)dx = \int_0^{\frac{\pi}{4}} \sin(2x)dx - \int_0^{\frac{\pi}{4}} \cos^2(2x)\sin(2x)dx = \frac{1}{2} - \frac{1}{6} = \frac{1}{3}$$

12.2.21. Problem 12.21

Calculate the definite integral:

$$\int_0^{\frac{\pi}{2}} x(1 + \cos x)dx$$

The domain of the integrand function is ***R***.

The integration interval $\left[0, \frac{\pi}{2}\right]$ is within the domain of the integrand function on which the function is continuous, so the function is integrable on the interval.

Considering that the integral on a bounded interval of a sum is equal to the sum of the integrals (Eq. 8.12), we can write:

$$\int_0^{\frac{\pi}{2}} x(1 + \cos x)dx = \int_0^{\frac{\pi}{2}} (x + x\cos x)dx = \int_0^{\frac{\pi}{2}} xdx + \int_0^{\frac{\pi}{2}} x\cos xdx$$

The first integral is:

$$\int_0^{\frac{\pi}{2}} x dx = \left[\frac{x^2}{2}\right]_0^{\frac{\pi}{2}} = \frac{1}{2}\left[\frac{\pi^2}{4} - 0\right] = \frac{\pi^2}{8}$$

The second integral can be calculated by parts.

$$\int f(x) \cdot g(x)dx = f(x) \cdot G(x) - \int f'(x) \cdot G(x)dx$$

We assume that:

$$f(x) = x \quad \rightarrow \quad f'(x) = 1$$
$$g(x) = G'(x) = \cos x \quad \rightarrow \quad G(x) = \sin x$$

We have:

$$\int_0^{\frac{\pi}{2}} x\cos x dx = [x\sin x]_0^{\frac{\pi}{2}} - \int_0^{\frac{\pi}{2}} \sin x dx = [x\sin x]_0^{\frac{\pi}{2}} - [-\cos x]_0^{\frac{\pi}{2}} =$$
$$= \left[\frac{\pi}{2}\sin\frac{\pi}{2} - 0\right] + \left[\cos\frac{\pi}{2} - \cos 0\right] = \frac{\pi}{2} - 1$$

We obtain therefore:

$$\int_0^{\frac{\pi}{2}} x(1 + \cos x)dx = \frac{\pi^2}{8} + \frac{\pi}{2} - 1$$

12.2.22. Problem 12.22

Calculate the definite integral:

$$\int_0^1 \frac{\log(x+1)}{x+1} dx$$

The domain of the integrand is given by the intersection of the function domains in the numerator and the denominator, minus the values in which the denominator is equal to 0.

The logarithm function is defined by:

$$x+1>0 \quad \Rightarrow \quad x>-1$$

so the domain of the numerator is:

$$D=]-1,+\infty[$$

The denominator (its domain is $\boldsymbol{R}$) equals to 0 for:

$$x+1=0 \quad \Rightarrow \quad x=-1$$

value already excluded from the domain of the numerator.

The function is therefore defined by:

$$f:]-1,+\infty[\rightarrow \boldsymbol{R}$$

The integration interval $[0,1]$ is within the domain of the integrand function on which the function is continuous, so the function is integrable on the interval.

The integral can be calculated by substitution:

$$t=\log(x+1) \quad \Rightarrow \quad dt=\frac{dt}{dx}dx=\frac{d}{dx}[\log(x+1)]dx=\frac{1}{x+1}dx$$

The limits of integration are:

$$x=0 \quad \rightarrow \quad t_{\inf}=\log(0+1)=\log 1=0$$

$$x=1 \quad \rightarrow \quad t_{\sup}=\log(1+1)=\log 2$$

The integral becomes:

$$\int_0^1 \frac{\log(x+1)}{x+1}dx=\int_0^{\log 2} t\,dt$$

A primitive of the function:

$$f(t) = F'(t) = t$$

it is:

$$F(t) = \frac{1}{2}t^2$$

We obtain:

$$\int_0^{\log 2} t dt = \left[\frac{1}{2}t^2\right]_0^{\log 2} = \frac{1}{2}(\log 2)^2 \equiv \frac{1}{2}\log^2 2$$

The integral is therefore:

$$\int_0^1 \frac{\log(x+1)}{x+1} dx = \frac{1}{2}\log^2 2$$

12.2.23. Problem 12.23

Calculate the definite integral:

$$\int_0^1 (x+1)^2 \cdot \log(x+1) dx$$

The domain of the integrand is determined by the domain of the logarithm function whose argument must be:

$$x+1>0 \quad \Rightarrow \quad x>-1$$

so the domain is:

$$D = \left]-1,+\infty\right[$$

The integration interval $[0,1]$ is within the domain of the integrand function on which the function is continuous, so the function is integrable on the interval.

The integral can be calculated by parts:

$$\int f(x)\cdot g(x)dx = f(x)\cdot G(x) - \int f'(x)\cdot G(x)dx$$

We assume that:

$$f(x) = \log(x+1) \quad , \quad g(x) = (x+1)^2$$
$$f'(x) = \frac{1}{x+1} \quad , \quad G(x) = \frac{1}{3}(x+1)^3$$

The definite integral is given by:

$$\int_0^1 (x+1)^2 \cdot \log(x+1)dx =$$
$$= \left[\log(x+1)\cdot\frac{1}{3}(x+1)^3\right]_0^1 - \int_0^1 \frac{1}{x+1}\cdot\frac{1}{3}(x+1)^3 dx =$$
$$= \left[\log(x+1)\cdot\frac{1}{3}(x+1)^3\right]_0^1 - \int_0^1 \frac{1}{3}(x+1)^2 dx =$$
$$= \left[\log(x+1)\cdot\frac{1}{3}(x+1)^3\right]_0^1 - \left[\frac{1}{9}(x+1)^3\right]_0^1 =$$
$$= \log(1+1)\cdot\frac{1}{3}(1+1)^3 - \log(0+1)\cdot\frac{1}{3}(0+1)^3 - \left[\frac{1}{9}(1+1)^3 - \frac{1}{9}(0+1)^3\right] =$$
$$= \frac{2^3}{3}\log 2 - \frac{1}{3}\log 1 - \frac{2^3}{9} + \frac{1}{9} = \frac{8}{3}\log 2 - \frac{8}{9} + \frac{1}{9} = \frac{8}{3}\log 2 - \frac{7}{9}$$

The integral is therefore:

$$\int_0^1 (x+1)^2 \cdot \log(x+1)dx = \frac{8}{3}\log 2 - \frac{7}{9}$$

12.2.24. Problem 12.24

Calculate the definite integral:

$$\int_{-\frac{\pi}{2}}^{\frac{\pi}{2}} \cos(x+1)\cdot \sin^3(x+1)dx$$

The domain of the integrand function is ***R***.

The integration interval $\left[-\frac{\pi}{2},\frac{\pi}{2}\right]$ is within the domain of the integrand function on which the function is continuous, so the function is integrable on the interval.

The integral can be calculated by substitution:

$$t = \sin(x+1)$$

$$dt = \frac{dt}{dx}dx = \frac{d}{dx}[\sin(x+1)]dx = \cos(x+1)dx$$

The limits of integration are:

$$x = -\frac{\pi}{2} \quad \to \quad t_{\inf} = \sin\left(-\frac{\pi}{2}+1\right)$$

$$x = \frac{\pi}{2} \quad \to \quad t_{\sup} = \sin\left(\frac{\pi}{2}+1\right)$$

The definite integral is given by:

$$\int_{-\frac{\pi}{2}}^{\frac{\pi}{2}} \cos(x+1)\cdot \sin^3(x+1)dx = \int_{\sin\left(-\frac{\pi}{2}+1\right)}^{\sin\left(\frac{\pi}{2}+1\right)} t^3 dt$$

We obtain:

$$\int_{\sin\left(-\frac{\pi}{2}+1\right)}^{\sin\left(\frac{\pi}{2}+1\right)} t^3 dt = \left[\frac{1}{4}t^4\right]_{\sin\left(-\frac{\pi}{2}+1\right)}^{\sin\left(\frac{\pi}{2}+1\right)} = \frac{1}{4}\sin^4\left(\frac{\pi}{2}+1\right) - \frac{1}{4}\sin^4\left(-\frac{\pi}{2}+1\right)$$

The integral is therefore:

$$\int_{-\frac{\pi}{2}}^{\frac{\pi}{2}} \cos(x+1)\cdot\sin^3(x+1)dx = \frac{1}{4}\sin^4\left(\frac{\pi}{2}+1\right) - \frac{1}{4}\sin^4\left(-\frac{\pi}{2}+1\right)$$

12.2.25. Problem 12.25

Calculate the definite integral:

$$\int_{0}^{\frac{\pi}{2}} \cos x \cdot e^{(\sin x+1)} dx$$

The domain of the integrand function is ***R***.

The integration interval $\left[0,\frac{\pi}{2}\right]$ is within the domain of the integrand function on which the function is continuous, so the function is integrable on the interval.

The integral can be calculated by substitution:

$$t = \sin x + 1 \quad \Rightarrow \quad dt = \frac{dt}{dx}dx = \frac{d}{dx}(\sin x + 1)dx = \cos x dx$$

The limits of integration are:

$$x = 0 \quad \rightarrow \quad t_{\text{inf}} = \sin 0 + 1 = 0 + 1 = 1$$

$$x = \frac{\pi}{2} \quad \rightarrow \quad t_{\text{sup}} = \sin\frac{\pi}{2} + 1 = 1 + 1 = 2$$

The integral is therefore:

$$\int_{0}^{\frac{\pi}{2}} \cos x \cdot e^{(\sin x+1)} dx = \int_{1}^{2} e^t dt = \left[e^t\right]_1^2 = e^2 - e = e\cdot(e-1)$$

12.2.26. Problem 12.26

Calculate the definite integral:

$$\int_0^{\frac{\pi}{2}} \sin x \cdot e^x dx$$

The domain of the integrand function is ***R***.

The integration interval $\left[0, \frac{\pi}{2}\right]$ is within the domain of the integrand function on which the function is continuous, so the function is integrable on the interval.

The integral can be calculated by parts:

$$\int f(x) \cdot g(x) dx = f(x) \cdot G(x) - \int f'(x) \cdot G(x) dx$$

We assume that:

$$f(x) = \sin x \quad , \quad g(x) = e^x$$
$$f'(x) = \cos x \quad , \quad G(x) = e^x$$

The definite integral is given by:

$$\int_0^{\frac{\pi}{2}} \sin x \cdot e^x dx = \left[\sin x \cdot e^x\right]_0^{\frac{\pi}{2}} - \int_0^{\frac{\pi}{2}} \cos x \cdot e^x dx$$

The integral:

$$\int_0^{\frac{\pi}{2}} \cos x \cdot e^x dx$$

It can in turn be calculated by parts, assuming that:

$$f(x) = \cos x \quad , \quad g(x) = e^x$$
$$f'(x) = -\sin x \quad , \quad G(x) = e^x$$

We have:

$$\int_0^{\frac{\pi}{2}} \cos x \cdot e^x dx = \left[\cos x \cdot e^x\right]_0^{\frac{\pi}{2}} - \int_0^{\frac{\pi}{2}} -\sin x \cdot e^x dx = \left[\cos x \cdot e^x\right]_0^{\frac{\pi}{2}} + \int_0^{\frac{\pi}{2}} \sin x \cdot e^x dx$$

Returning to the initial integral, we have:

$$\int_0^{\frac{\pi}{2}} \sin x \cdot e^x dx = \left[\sin x \cdot e^x\right]_0^{\frac{\pi}{2}} - \int_0^{\frac{\pi}{2}} \cos x \cdot e^x dx =$$

$$= \left[\sin x \cdot e^x\right]_0^{\frac{\pi}{2}} - \left[\cos x \cdot e^x\right]_0^{\frac{\pi}{2}} - \int_0^{\frac{\pi}{2}} \sin x \cdot e^x dx$$

We obtain thus:

$$\int_0^{\frac{\pi}{2}} \sin x \cdot e^x dx + \int_0^{\frac{\pi}{2}} \sin x \cdot e^x dx = 2\int_0^{\frac{\pi}{2}} \sin x \cdot e^x dx = \left[\sin x \cdot e^x\right]_0^{\frac{\pi}{2}} - \left[\cos x \cdot e^x\right]_0^{\frac{\pi}{2}}$$

$$\int_0^{\frac{\pi}{2}} \sin x \cdot e^x dx = \frac{1}{2}\left[\sin\frac{\pi}{2} \cdot e^{\frac{\pi}{2}} - \sin 0 \cdot e^0 - \cos\frac{\pi}{2} \cdot e^{\frac{\pi}{2}} + \cos 0 \cdot e^0\right] =$$

$$= \frac{1}{2}\left[1 \cdot e^{\frac{\pi}{2}} - 0 \cdot e^0 - 0 \cdot e^{\frac{\pi}{2}} + 1 \cdot e^0\right] = \frac{1}{2}\left[e^{\frac{\pi}{2}} + 1\right]$$

The integral is therefore:

$$\int_0^{\frac{\pi}{2}} \sin x \cdot e^x dx = \frac{1}{2}\left(e^{\frac{\pi}{2}} + 1\right)$$

12.2.27. Problem 12.27

Calculate the definite integral:

$$\int_1^4 x^{\frac{1}{2}} \cdot \log x dx$$

The domain of the integrand is determined by the logarithm function and it is $\boldsymbol{R}^+$.

The integration interval $[1,4]$ is within the domain of the integrand function on which the function is continuous, so the function is integrable on the interval.

The integral can be calculated by parts:

$$\int f(x)\cdot g(x)dx = f(x)\cdot G(x) - \int f'(x)\cdot G(x)dx$$

We assume that:

$$f(x) = \log x \quad , \quad g(x) = x^{\frac{1}{2}}$$

$$f'(x) = \frac{1}{x} \quad , \quad G(x) = \frac{2}{3}x^{\frac{3}{2}}$$

The definite integral is given by:

$$\int_1^4 x^{\frac{1}{2}}\cdot \log x dx = \left[\log x\cdot \frac{2}{3}x^{\frac{3}{2}}\right]_1^4 - \int_1^4 \frac{1}{x}\cdot\frac{2}{3}x^{\frac{3}{2}}dx =$$

$$= \left[\log x\cdot \frac{2}{3}x^{\frac{3}{2}}\right]_1^4 - \int_1^4 \frac{2}{3}x^{\frac{3}{2}-1}dx = \left[\log x\cdot \frac{2}{3}x^{\frac{3}{2}}\right]_1^4 - \int_1^4 \frac{2}{3}x^{\frac{1}{2}}dx =$$

$$= \left[\log x\cdot \frac{2}{3}x^{\frac{3}{2}}\right]_1^4 - \left[\frac{2}{3}\cdot\frac{2}{3}x^{\frac{3}{2}}\right]_1^4 = \left[\log x\cdot \frac{2}{3}x^{\frac{3}{2}}\right]_1^4 - \left[\frac{4}{9}x^{\frac{3}{2}}\right]_1^4 =$$

$$= \left[\log 4\cdot\frac{2}{3}4^{\frac{3}{2}} - \log 1\cdot\frac{2}{3}1^{\frac{3}{2}}\right] - \left[\frac{4}{9}4^{\frac{3}{2}} - \frac{4}{9}1^{\frac{3}{2}}\right] = \frac{16}{3}\log 4 - \frac{32}{9} + \frac{4}{9} =$$

$$= \frac{16}{3}\log 4 - \frac{28}{9} = \frac{16}{3}\log 2^2 - \frac{28}{9} = \frac{16}{3}\cdot 2\log 2 - \frac{28}{9} = \frac{32}{3}\log 2 - \frac{28}{9}$$

The integral is therefore:

$$\int_1^4 x^{\frac{1}{2}}\cdot \log x dx = \frac{32}{3}\log 2 - \frac{28}{9}$$

12.2.28. Problem 12.28

Calculate the definite integral:

$$\int_0^1 x^2 \cdot e^x dx$$

The domain of the integrand function is ***R***.

The integration interval $[0,1]$ is within the domain of the integrand function on which the function is continuous, so the function is integrable on the interval.

The integral can be calculated by parts:

$$\int f(x) \cdot g(x) dx = f(x) \cdot G(x) - \int f'(x) \cdot G(x) dx$$

We assume that:

$$f(x) = x^2 \quad , \quad g(x) = e^x$$
$$f'(x) = 2x \quad , \quad G(x) = e^x$$

The definite integral is given by:

$$\int_0^1 x^2 \cdot e^x dx = \left[x^2 \cdot e^x\right]_0^1 - \int_0^1 2x \cdot e^x dx$$

The integral:

$$\int_0^1 2x \cdot e^x dx$$

It can in turn be calculated by parts, assuming that:

$$f(x) = 2x \quad , \quad g(x) = e^x$$
$$f'(x) = 2 \quad , \quad G(x) = e^x$$

We have:

$$\int_0^1 2x \cdot e^x dx = \left[2x \cdot e^x\right]_0^1 - \int_0^1 2e^x dx = \left[2x \cdot e^x\right]_0^1 - \left[2e^x\right]_0^1$$

Returning to the initial integral, we have:

$$\int_0^1 x^2 \cdot e^x dx = \left[x^2 \cdot e^x\right]_0^1 - \int_0^1 2x \cdot e^x dx =$$

$$= \left[x^2 \cdot e^x\right]_0^1 - \left[2x \cdot e^x\right]_0^1 + \int_0^1 2e^x dx = \left[x^2 \cdot e^x\right]_0^1 - \left[2x \cdot e^x\right]_0^1 + \left[2e^x\right]_0^1 =$$

$$= \left[e - 0 \cdot e^0\right] - \left[2e - 0 \cdot e^0\right] + \left[2e - 2e^0\right] = e - 2e + 2e - 2 = e - 2$$

The integral is therefore:

$$\int_0^1 x^2 \cdot e^x dx = e - 2$$

12.2.29. Problem 29

Calculate the integral:

$$\int_0^1 \log x dx$$

For the logarithm function it holds:

$$f(x) = \log x \ , \ f : \boldsymbol{R}^+ \to \boldsymbol{R}$$

$$\lim_{x \to 0+} \log x = -\infty$$

In the case under consideration, we want to calculate the integral of an unbounded function on a bounded integration interval.

It is an improper integral that can be calculated according to the following method (Eq. 8.32):

$$\exists \lim_{\alpha\to 0+} \int_{\alpha}^{1} \log x dx \in \boldsymbol{R} \quad \Rightarrow \quad \int_{0}^{1} \log x dx = \lim_{\alpha\to 0+} \int_{\alpha}^{1} \log x dx$$

It must be calculated the integral:

$$\int_{\alpha}^{1} \log x dx$$

The integral can be calculated by parts, as was done previously for the indefinite integral (Eq. 7.79, sub-subsection 7.5.1.3):

$$\int \log x dx = x(\log x - 1) + c$$

The definite integral is therefore:

$$\int_{\alpha}^{1} \log x dx = [x(\log x - 1)]_{\alpha}^{1}$$

We have:

$$\lim_{\alpha\to 0+} \int_{\alpha}^{1} \log x dx = \lim_{\alpha\to 0+} [x(\log x - 1)]_{\alpha}^{1} =$$
$$= \lim_{\alpha\to 0+} [1\cdot(\log 1 - 1) - \alpha\cdot(\log\alpha - 1)] =$$
$$= 1\cdot(\log 1 - 1) - \lim_{\alpha\to 0+} [\alpha\cdot(\log\alpha - 1)]$$

It holds:

$$1\cdot(\log 1 - 1) = -1$$

and:

$$\lim_{\alpha\to 0+} [\alpha\cdot(\log\alpha - 1)] = \lim_{\alpha\to 0+} \alpha\log\alpha - \lim_{\alpha\to 0+} \alpha = \lim_{\alpha\to 0+} \alpha\log\alpha$$

The limit is:

$$\lim_{\alpha\to 0+}\alpha\log\alpha = \lim_{\alpha\to 0+}\frac{\log\alpha}{\alpha^{-1}} = \frac{-\infty}{+\infty}$$

It is an indeterminate form, to solve it we can apply De L'Hôpital's rule:

$$\lim_{\alpha\to 0+}\frac{\log\alpha}{\alpha^{-1}} = \lim_{\alpha\to 0+}\frac{\frac{d}{d\alpha}(\log\alpha)}{\frac{d}{d\alpha}\left(\alpha^{-1}\right)} = \lim_{\alpha\to 0+}\frac{\frac{1}{\alpha}}{-\alpha^{-2}} =$$

$$= \lim_{\alpha\to 0+}\left(-\frac{1}{\alpha^{-1}}\right) = -\lim_{\alpha\to 0+}\alpha = 0$$

The integral is therefore:

$$\int_0^1 \log x dx = \lim_{\alpha\to 0+}\int_\alpha^1 \log x dx = -1$$

12.2.30. Problem 30

Calculate the integral:

$$\int_1^{+\infty}\frac{1}{x^2}dx$$

For the integrand function applies:

$$f(x) = \frac{1}{x^2}\ ,\ f:\boldsymbol{R}-\{0\}\to\boldsymbol{R}$$

$$\lim_{x\to+\infty}\frac{1}{x^2} = 0$$

The function is bounded on the integration interval, so we fall back in the case of an integral of a bounded function on an unbounded interval.

It is an improper integral that can be calculated according to the following method (Eq. 8.39):

$$\exists \lim_{\beta \to +\infty} \int_1^{\beta} \frac{1}{x^2}\,dx \in \boldsymbol{R} \quad \Rightarrow \quad \int_1^{+\infty} \frac{1}{x^2}\,dx = \lim_{\beta \to +\infty} \int_1^{\beta} \frac{1}{x^2}\,dx$$

We have:

$$\int_1^{\beta} \frac{1}{x^2}\,dx = \int_1^{\beta} x^{-2}\,dx = \left[-x^{-1}\right]_1^{\beta} = -\beta^{-1} - (-1) = -\beta^{-1} + 1$$

The improper integral is therefore:

$$\int_1^{+\infty} \frac{1}{x^2}\,dx = \lim_{\beta \to +\infty} \int_1^{\beta} \frac{1}{x^2}\,dx = \lim_{\beta \to +\infty} \left(-\beta^{-1} + 1\right) =$$

$$= \lim_{\beta \to +\infty} \left(-\frac{1}{\beta} + 1\right) = 1$$

12.2.31. Problem 31

Calculate the integral:

$$\int_0^{\frac{1}{2}} \frac{dx}{\sqrt{1-4x^2}}$$

The domain of the integrand function is given by the values for which the argument of the square root function being in the denominator is > 0. In the specific case by the values:

$$1-4x^2 > 0 \quad \Rightarrow \quad 4x^2 < 1 \quad \Rightarrow \quad |x| < \frac{1}{2}$$

For the integrand function applies:

$$f(x) = \frac{1}{\sqrt{1-4x^2}} \ , \ f : \left]-\frac{1}{2}, \frac{1}{2}\right[\to \boldsymbol{R}$$

$$\lim_{x \to \frac{1}{2}^-} \frac{1}{\sqrt{1-4x^2}} = +\infty$$

In the case under consideration, we want to calculate the integral of an unbounded function on a bounded integration interval.

It is an improper integral that can be calculated according to the following method (Eq. 8.35):

$$\exists \lim_{\beta \to \frac{1}{2}^-} \int_0^{\beta} \frac{dx}{\sqrt{1-4x^2}} \in \boldsymbol{R} \quad \Rightarrow \quad \int_0^{\frac{1}{2}} \frac{dx}{\sqrt{1-4x^2}} = \lim_{\beta \to \frac{1}{2}^-} \int_0^{\beta} \frac{dx}{\sqrt{1-4x^2}}$$

In this specific case, it can be verified that it is possible to proceed even without calculating the limit.

The integral does not seem to admit a primitive, nor does it appear to be integrable by parts, so it can be assumed to use the method of integration by substitution.

To identify the substitution of the most suitable variable for the calculation of the integral, it can be observed that the denominator is:

$$\sqrt{1-4x^2}$$

that is, the square root of the difference between 1 and the square of $2x$.

Considering that for trigonometric functions it holds:

$$\sin^2 t + \cos^2 t = 1$$

From which, for example:

$$\cos^2 t = 1 - \sin^2 t$$

We can do variable substitution:

$$2x = \sin t \quad \rightarrow \quad x = \frac{1}{2}\sin t$$

$$dx = \frac{dx}{dt}dt = \frac{d}{dt}\left(\frac{1}{2}\sin t\right)dt = \frac{1}{2}\cos t dt$$

The limits of integration are:

$$x = 0 \quad \Rightarrow \quad \frac{1}{2}\sin t = 0 \quad \Rightarrow \quad t = 0$$

$$x = \frac{1}{2} \quad \Rightarrow \quad \frac{1}{2}\sin t = \frac{1}{2} \quad \Rightarrow \quad t = \frac{\pi}{2}$$

We obtain:

$$\int_0^{\frac{1}{2}} \frac{dx}{\sqrt{1-4x^2}} = \int_0^{\frac{\pi}{2}} \frac{\frac{1}{2}\cos t dt}{\sqrt{1-\sin^2 t}} = \frac{1}{2}\int_0^{\frac{\pi}{2}} \frac{\cos t dt}{\sqrt{1-\sin^2 t}} = \frac{1}{2}\int_0^{\frac{\pi}{2}} \frac{\cos t dt}{\sqrt{\cos^2 t}} =$$

$$= \frac{1}{2}\int_0^{\frac{\pi}{2}} \frac{\cos t}{\cos t} dt = \frac{1}{2}\int_0^{\frac{\pi}{2}} dt = \frac{1}{2}[t]_0^{\frac{\pi}{2}} = \frac{\pi}{4}$$

The integral is therefore:

$$\int_0^{\frac{1}{2}} \frac{dx}{\sqrt{1-4x^2}} = \frac{\pi}{4}$$

It is not necessary to calculate the limit because with the substitution of the variable carried out the integrand function is:

$$\frac{\cos t}{\cos t} = 1$$

12.2.32. Problem 32

Calculate the integral:

$$\int_0^{+\infty} e^{-\frac{x}{3}} dx$$

The integral falls in the case of unbounded integration interval and bounded function. In fact, the upper limit of integration is $+\infty$, while for the function it holds:

$$f(x) = e^{-\frac{x}{3}}\,,\ f : \boldsymbol{R} \to \boldsymbol{R}, \qquad \lim_{x \to +\infty} e^{-\frac{x}{3}} = 0, \qquad f \in C_{\boldsymbol{R}}$$

The improper integral is given by:

$$\int_0^{+\infty} e^{-\frac{x}{3}} dx = \lim_{\beta \to +\infty} \int_0^{\beta} e^{-\frac{x}{3}} dx$$

The integrand function admits primitive:

$$\int e^{-\frac{x}{3}} dx = -3e^{-\frac{x}{3}} + c$$

Thus:

$$\int_0^{+\infty} e^{-\frac{x}{3}} dx = \lim_{\beta \to +\infty} \int_0^{\beta} e^{-\frac{x}{3}} dx = \lim_{\beta \to +\infty} \left[-3e^{-\frac{x}{3}} \right]_0^{\beta} =$$

$$= \lim_{\beta \to +\infty} \left(-3e^{-\frac{\beta}{3}} \right) - \left(-3e^{-\frac{0}{3}} \right) = 0 + 3 = 3$$

The integral is therefore:

$$\int_0^{+\infty} e^{-\frac{x}{3}} dx = \lim_{\beta \to +\infty} \int_0^{\beta} e^{-\frac{x}{3}} dx = 3$$

12.2.33. Problem 33

Calculate the integral:

$$\int_{-\infty}^{0} e^{2x} dx$$

The integral falls in the case of unbounded integration interval and bounded function. In fact, the lower limit of integration is $-\infty$, while for the function it holds:

$$f(x)=e^{2x}, \qquad D=\boldsymbol{R}, \qquad \lim_{x\to-\infty} e^{2x}=0, \qquad f\in C_{\boldsymbol{R}}$$

The improper integral is given by:

$$\int_{-\infty}^{0} e^{2x}dx = \lim_{\alpha\to-\infty}\int_{\alpha}^{0} e^{2x}dx$$

The integrand function admits primitive:

$$f(x)=e^{2x} \quad , \quad F(x)=\frac{1}{2}e^{2x}$$

Thus:

$$\int_{-\infty}^{0} e^{2x}dx = \lim_{\alpha\to-\infty}\int_{\alpha}^{0} e^{2x}dx = \lim_{\alpha\to-\infty}\left[\frac{1}{2}e^{2x}\right]_{\alpha}^{0} = \lim_{\alpha\to-\infty}\left[\frac{1}{2}e^{0}-\frac{1}{2}e^{2\alpha}\right]=$$

$$=\frac{1}{2}-\frac{1}{2}e^{-\infty}=\frac{1}{2}-0=\frac{1}{2}$$

The integral is therefore:

$$\int_{-\infty}^{0} e^{2x}dx = \lim_{\alpha\to-\infty}\int_{\alpha}^{0} e^{2x}dx = \frac{1}{2}$$

12.2.34. Problem 34

Calculate the integral:

$$\int_{0}^{+\infty} e^{-\left(\frac{x+1}{4}\right)}dx$$

The integral falls in the case of unbounded integration interval and bounded function. In fact, the upper limit of integration is $+\infty$, while for the function it holds:

$$f(x) = e^{-\left(\frac{x+1}{4}\right)}, \qquad D = \boldsymbol{R}, \qquad \lim_{x\to+\infty} e^{-\left(\frac{x+1}{4}\right)} = 0, \qquad f \in C_R$$

The improper integral is given by:

$$\int_0^{+\infty} e^{-\left(\frac{x+1}{4}\right)} dx = \lim_{\beta\to+\infty} \int_0^{\beta} e^{-\left(\frac{x+1}{4}\right)} dx$$

The integrand function admits primitive:

$$f(x) = e^{-\left(\frac{x+1}{4}\right)} \qquad , \qquad F(x) = -4e^{-\left(\frac{x+1}{4}\right)}$$

Thus:

$$\int_0^{+\infty} e^{-\left(\frac{x+1}{4}\right)} dx = \lim_{\beta\to+\infty} \int_0^{\beta} e^{-\left(\frac{x+1}{4}\right)} dx = \lim_{\beta\to+\infty} \left[-4e^{-\left(\frac{x+1}{4}\right)}\right]_0^{\beta} =$$

$$= -4 \lim_{\beta\to+\infty} \left[e^{-\left(\frac{\beta+1}{4}\right)} - e^{-\left(\frac{0+1}{4}\right)} \right] = -4e^{-\infty} + 4e^{-\frac{1}{4}} = 4e^{-\frac{1}{4}}$$

The integral is therefore:

$$\int_0^{+\infty} e^{-\left(\frac{x+1}{4}\right)} dx = \lim_{\beta\to+\infty} \int_0^{\beta} e^{-\left(\frac{x+1}{4}\right)} dx = 4e^{-\frac{1}{4}}$$

12.2.35. Problem 35

Calculate the definite integral:

$$\int_0^{+\infty} xe^{-x^2} dx$$

The integral falls in the case of unbounded integration interval and bounded function. In fact, the upper limit of integration is $+\infty$, while for the function it holds:

$$\lim_{x\to+\infty} xe^{-x^2} = +\infty \cdot 0$$

It is an indeterminate form, the limit can be written as:

$$\lim_{x\to+\infty} \frac{x}{e^{x^2}} = \frac{+\infty}{+\infty}$$

It is still an indeterminate form, however written in this way it allows to apply De L'Hôpital's rule:

$$\lim_{x\to+\infty} \frac{x}{e^{x^2}} = \lim_{x\to+\infty} \frac{\frac{d}{dx}(x)}{\frac{d}{dx}\left(e^{x^2}\right)} = \lim_{x\to+\infty} \frac{1}{2xe^{x^2}} = \frac{1}{(+\infty)\cdot(+\infty)} = 0$$

Thus:

$$\lim_{x\to+\infty} xe^{-x^2} = 0$$

$$f(x) = xe^{-x^2} \quad D = \boldsymbol{R}, \qquad \lim_{x\to+\infty} xe^{-x^2} = 0, \qquad f \in C_{\boldsymbol{R}}$$

The improper integral is given by:

$$\int_0^{+\infty} xe^{-x^2} dx = \lim_{\beta\to+\infty} \int_0^{\beta} xe^{-x^2} dx$$

The integral can be calculated by substitution:

$$t = x^2$$

$$dt = \frac{dt}{dx} dx = \frac{d}{dx}\left(x^2\right)dx = 2xdx \qquad \Rightarrow \qquad xdx = \frac{1}{2} dt$$

The limits of integration are:

$$x = 0 \quad \rightarrow \quad t_{\text{inf}} = 0^2 = 0$$

$$x = \beta \quad \rightarrow \quad t_{\text{sup}} = \beta^2$$

We have:

$$\int_0^{+\infty} xe^{-x^2}dx = \lim_{\beta\to+\infty}\int_0^{\beta} xe^{-x^2}dx = \lim_{\beta\to+\infty}\int_0^{\beta^2}\frac{1}{2}e^{-t}dt = \lim_{\beta\to+\infty}\left[-\frac{1}{2}e^{-t}\right]_0^{\beta^2} =$$

$$= \lim_{\beta\to+\infty}\left[-\frac{1}{2}e^{-\beta^2} - \left(-\frac{1}{2}\right)e^{-0}\right] = \lim_{\beta\to+\infty}\left[-\frac{1}{2}e^{-\beta^2} + \frac{1}{2}\right] =$$

$$= -\frac{1}{2}e^{-\infty} + \frac{1}{2} = 0 + \frac{1}{2} = \frac{1}{2}$$

The integral is therefore:

$$\int_0^{+\infty} xe^{-x^2}dx = \lim_{\beta\to+\infty}\int_0^{\beta} xe^{-x^2}dx = \frac{1}{2}$$

About the Author

Claudio Caprara is Associate Professor at the University of Bologna.

His main fields of research and teaching are: Agricultural process control and automation, precision agriculture, agricultural systems, logistics and production chains, biosystems engineering, renewable energy and GIS models.

He is the author/co-author of over 120 papers published in scientific journals and in congress proceedings.

INDEX

E

F

G

H

I

K

L

M

N

O

P

R

S

T

U

V

Y